DEVELOPMENTS IN HEAT EXCHANGER TECHNOLOGY—1

Edited by

D. CHISHOLM

Head of Applied Heat Transfer Division,
National Engineering Laboratory, Glasgow, UK

APPLIED SCIENCE PUBLISHERS LTD
LONDON

APPLIED SCIENCE PUBLISHERS LTD
RIPPLE ROAD, BARKING, ESSEX, ENGLAND

British Library Cataloguing in Publication Data

Developments in heat exchanger technology.—
(Developments series).
1
1. Heat exchangers
I. Chisholm, Duncan II. Series
621.4'022 TJ263

ISBN 0-85334-913-4

WITH 22 TABLES AND 86 ILLUSTRATIONS

Printed in Great Britain by Galliard (Printers) Ltd, Great Yarmouth

PREFACE

There must be very few industrial products which during some stage of their production are not raised to high temperature; heat exchanger technology is therefore pertinent, either directly or indirectly, to all branches of industry.

Developments in Heat Exchanger Technology—1 is, hopefully, the beginning of a series of volumes intended for use by the professional engineer in keeping himself abreast with developments in this field. Energy conservation in all its aspects has increased interest in heat transfer processes both in industry, transportation, and in the domestic situation. The emphasis throughout is on the heat exchanger and the thermal system, rather than on the heat exchange mechanism; in this book we have, for example, chapters on reboilers and condensers, rather than on reboiling and condensing.

This book concerns itself with the most common heat exchangers, an aspect I cover in more detail in Chapter 1. The final two chapters discuss two items particularly important in relation to energy conservation, the heat pump and waste heat recovery exchangers.

I was fortunate in that those internationally recognised authorities whom I approached to contribute to this book were willing and able to do so. I greatly appreciate their contributions and their alacrity in working to the relatively tight time schedules involved in the production of this book.

D. Chisholm
National Engineering Laboratory,
East Kilbride

CONTENTS

LIST OF CONTRIBUTORS

G. S. CATTELL

Process Development Department Manager, The A.P.V. Company Ltd, P.O. Box 4, Manor Royal, Crawley, Sussex RH10 2QB, UK.

D. CHISHOLM

Head, Applied Heat Transfer Division, National Engineering Laboratory, East Kilbride, Glasgow G75 0QU, UK.

P. FREUND

Head of the Heat Pump Section, Building Research Establishment, Building Research Station, Garston, Watford WD2 7JR, UK.

I. D. R. GRANT

Section Leader, Applied Heat Transfer Division, National Engineering Laboratory, East Kilbride, Glasgow G75 0QU, UK.

J. M. MCNAUGHT

Higher Scientific Officer, Applied Heat Transfer Division, National Engineering Laboratory, East Kilbride, Glasgow G75 0QU, UK.

C. D. R. NORTH

Technical Manager, GEA Airexchangers Ltd, Charles House, 7 Leicester Place, London WC2H 7BP, UK.

G. T. Polley

Section Leader, Applied Heat Transfer Division, National Engineering Laboratory, East Kilbride, Glasgow G75 0QU, UK.

D. A. Reay

Business Development Manager, International Research and Development Co. Ltd, Fossway, Newcastle upon Tyne NE6 2YD, UK.

J. D. Usher

Design and Development Manager, The A.P.V. Company Ltd, P.O. Box 4, Manor Royal, Crawley, Sussex RH10 2QB, UK.

D. R. Webb

Lecturer in Chemical Engineering, University of Manchester Institute of Science and Technology, Sackville Street, Manchester M60 1QD, UK.

R. L. Webb

Associate Professor, Department of Mechanical Engineering, The Pennsylvania State University, University Park, Pennsylvania 16802, USA.

Chapter 1

INTRODUCTION

D. CHISHOLM

National Engineering Laboratory, East Kilbride, Glasgow, UK

1. INTRODUCTION

The most general definition of a heat exchanger is a device for transferring heat from a heat source to a heat sink. To many professional engineers a more specific definition is *a device for transferring heat from one flowing fluid to another*; it is in this sense that the description *heat exchanger* is used here.

Table 1.1 serves to indicate the great variety of heat exchangers and their typical duties. It is not possible within a text of this size to discuss all these heat exchangers, and for that reason further sources of information are given in the bibliography at the end of this book; references are given there not only to literature on heat exchangers but also to literature on the various mechanisms of heat transfer. An index to the bibliography is also given.

1.1. Scope of Text

More than 90% of heat exchangers used in industry are of the shell-and-tube type, and for this reason three of the nine chapters are primarily concerned with this type of unit. Chapter 2 is concerned with shell-and-tube heat exchangers for single-phase applications; Chapter 3 includes a discussion of boiling in these devices, and Chapter 4 a discussion on condensation.

The most common heat exchanger of all is the automobile radiator, more than 10 million being produced per annum. The general description of a heat exchanger of that type is *compact heat exchanger*; these are discussed in Chapter 5.

TABLE 1.1

HEAT EXCHANGERS AND TYPICAL DUTIES

Heat exchanger	*Application*										
	Boiling	*Condensing*	*Corrosive*	*Fouling*	*Gas/gas*	*Gas/liquid*	*Gas-side enhancement*	*Liquid/liquid*	*Low residence time*	*Low temperature difference*	*Viscous fluid*
Air cooler		√				√					
Block type (graphite)			√					√			
Direct contact				√						√	
Double pipe									√		√
Evaporative coolers		√				√					
Embossed panel (heating jackets)						√		√			
Fluidised bed			√	√			√				
Finned tube (immersion coil)						√		√			
Hampson (coiled tubular)		√								√	
Heat pipe					√	√		√			
Lamella											
Packed bed							√				
Plate		√		√				√		√	
Plate-fin	√	√			√			√			
Plate spiral		√		√				√			
Rotary					√						
Scraped surface				√							√
Shell-and-tube	√	√						√		√	
Tube-in-plate						√					

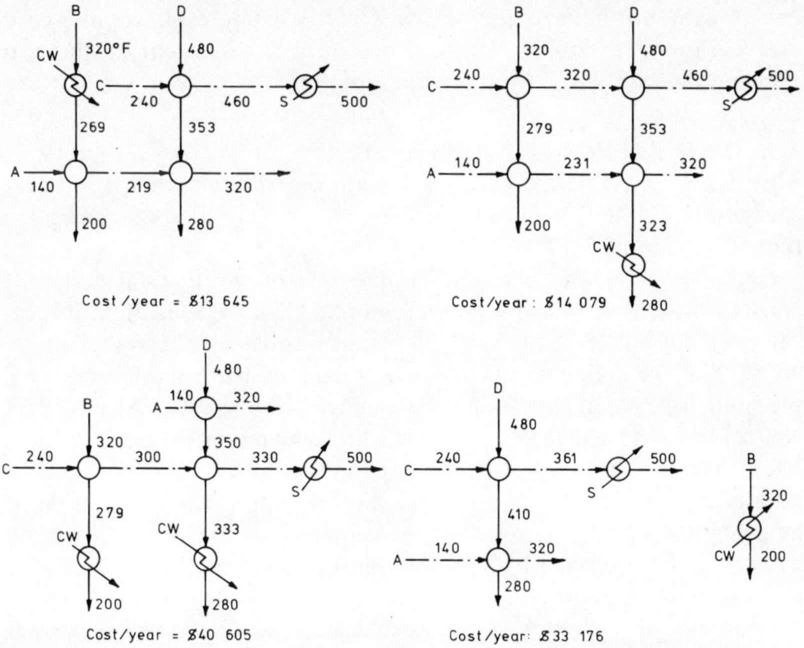

FIG. 1.1. Four alternative arrangements of heat exchangers.

Chapter 6 deals with the other major class of heat exchanger, the air cooler. These units are the heat sink of many industrial processes. In air coolers, on the air side, it is necessary to enhance the air-side heat transfer coefficient by means of extruded surfaces. Enhancement is of course relevant to all heat exchanger surfaces and is discussed in detail in Chapter 7.

The book concludes with discussion of two items particularly important in relation to energy conservation, the heat pump in Chapter 8, and exchangers for waste heat recovery in Chapter 9.

1.2. Heat Exchanger Networks

Many industrial processes involve a multitude of heat exchangers. Rathore and Powers[1] found that exchanging heat between two hot and two cold streams could be achieved in over 200 ways. Four possible arrangements are shown in Fig. 1.1. For the same initial and final flow stream conditions, a threefold difference in annual costs between 'best' and 'worst' methods was found.

Challand[2] has demonstrated how network optimisation of existing plant can result in substantial energy saving. Thus the capital costs, running cost, and energy utilisation of plant can be critically dependent on the heat exchanger network selection and optimisation.[3,4]

1.3. The Design Process: Current Status

This book is primarily concerned with the thermal design of heat exchangers. Some general comment on the industrial design process is therefore relevant.

Computer programs are now available[5,6] for the thermal design of conventional heat exchangers such as single-phase shell-and-tube exchangers, condensers, boilers, reboilers and air coolers. Programs generally are suitable for operation in the batch or demand batch mode, and some programs operate interactively, the engineer having the ability to gain control at intermediate points during the calculation process. In these interactive programs the scope of action available to the designer is normally limited at the present time. Too rigorous an interpretation of limits by the computer, resulting in commercially viable designs being discarded, is one reason why there is considerable interest in the industry for more interactive elements in programs.

Generally, the selection of a particular design is made at the thermal design stage, the thermal program carrying out the minimum of mechanical design necessary to enable this to be done. The output from the thermal design program is frequently made up of a number of designs from which the engineer makes his choice. The thermal design program will give approximate cost estimates; it is important at this stage that the costs of various designs are correct relatively, rather than in magnitude. The selection of the design which gives minimum capital cost is usually the extent of optimisation at the present time, though when the pump is integral with the exchanger, as in air coolers, pumping costs will also be considered.

1.4. The Design Process: Future Developments

The digital computer has revolutionised the heat exchanger design process during the last 10 years, and it is to be expected that during the next 15 years a major influence will continue to stem from that source. It is relevant therefore to look briefly at anticipated developments in computer technology. The following extract[7] gives an indication of expected trends:

> 'An impact of technology is to change the relative economics of the various aspects of computing. A comparison of costs in real terms now and in the future is likely to show the following pattern:

	1975	1980	1985	1990
peripherals	1·0	0·75	0·6	0·5
programming	1·0	1·25	1·4	1·5
processing	1·0	0·25	0·1	0·05
element/chip				
processors/k	1·0	15	100	500
storage/k	1·0	20	100	1 000

This pattern should be taken as no more than an indication of the trend; the cost of processing reflecting continued improvements in semi-conductor technology; the cost of peripherals decreasing slowly, limited by the relatively stable mechanical technology; whilst the cost of programming increases because of the greater demand for a scarce source.'

The vast anticipated changes in computer technology mentioned above should have a major influence on the manner in which the design process is handled in the future. It is to be expected that routine design will be carried out on desk-top computers (or personal computers as they are now being called), main frame machines being used for specialised design situations requiring large complex programs. To date, a factor limiting the use of highly sophisticated design methods has been the restrictions imposed by the size of computers; this will be increasingly less of a restriction as time goes on. The largest data bank currently existing will be available with the desk-top computer.

In the future the professional engineer will no doubt have his own personalised computer system, consisting of a personal computer with visual display, a floppy disc storage system, and printing facilities. He will know in detail the software under his control, and during his career he will have acquired a 'software library'.

1.5. Mechanical Design and Draughting

Mechanical design and draughting of shell-and-tube heat exchangers can now be carried out by computer[8,9] though at the present time the draughting facility is not extensively used in the heat exchanger industry. Software[10,11] suitable for general draughting application in the heat exchanger industry is now being examined by various organisations. Figure 1.2 shows a heat exchanger 'setting plan' produced by the computer program STEM.[8]

An important problem which is partly a mechanical and partly a thermal problem is that of tube vibration. It is essential in the case of tube banks to check the vibrational characteristics.[12]

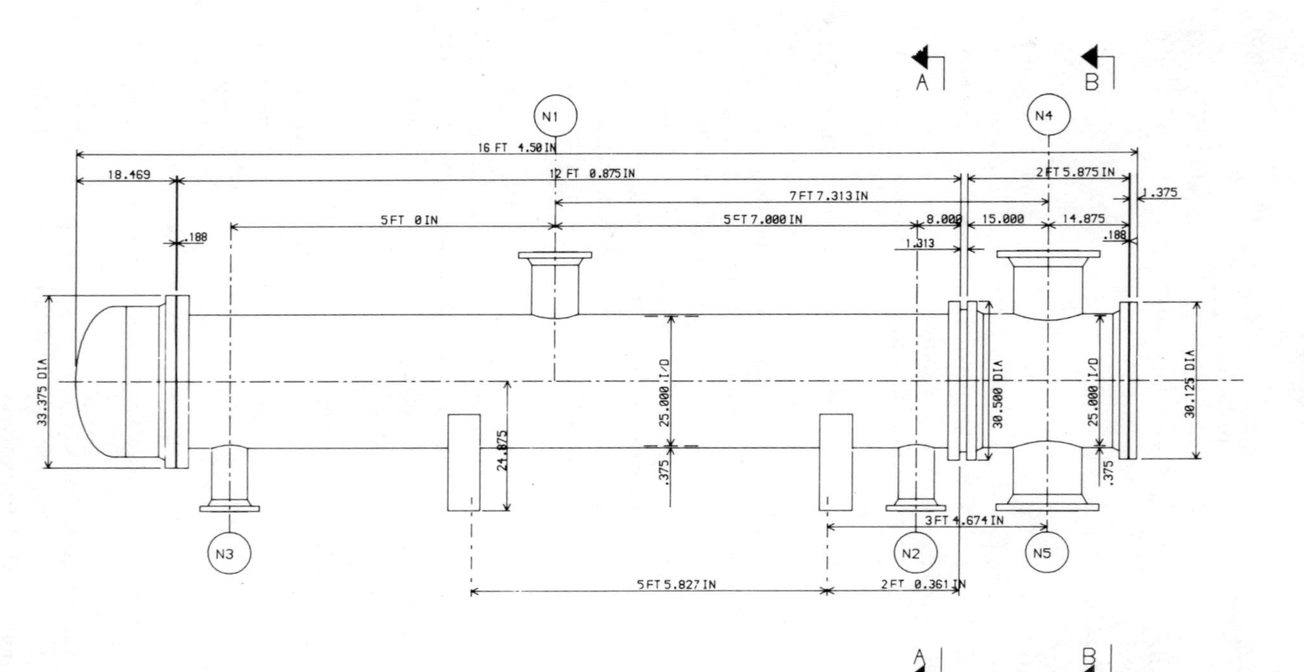

FIG. 1.2. Heat exchanger 'setting plan' produced by the computer program STEM.[8]

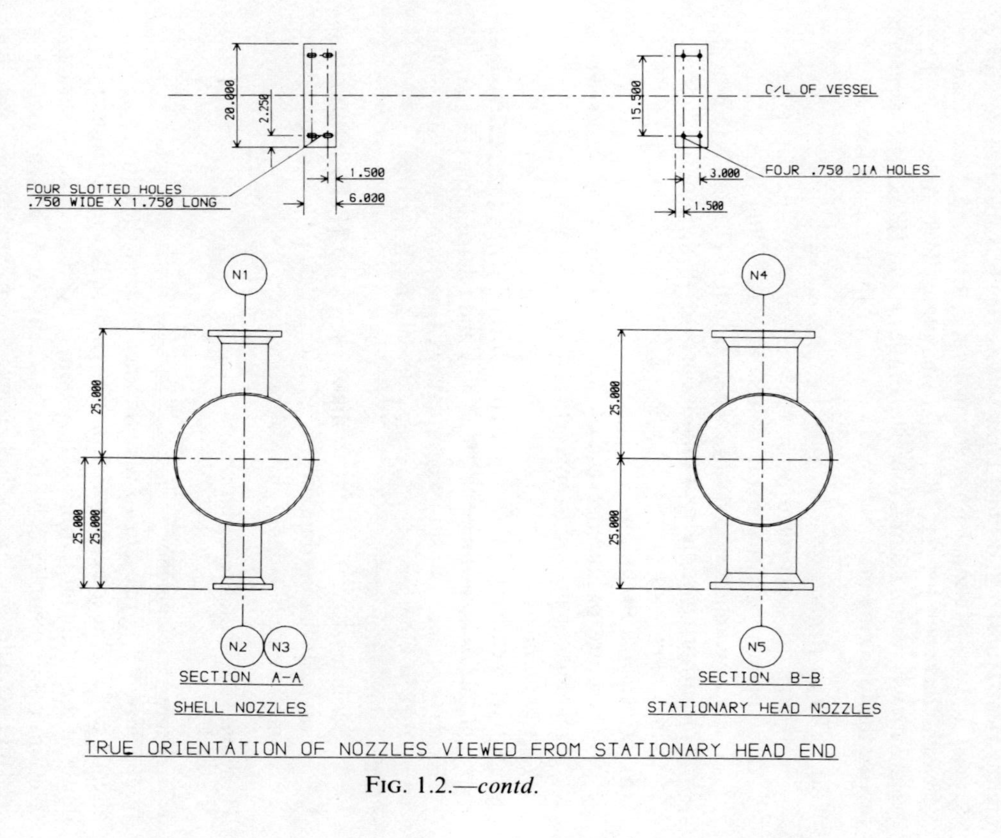
20.000
2.250
FOUR SLOTTED HOLES
.750 WIDE X 1.750 LONG
1.500
6.000
15.500
C/L OF VESSEL
FOJR .750 DIA HOLES
3.000
1.500
N1
N4
25.000
25.000
25.000
25.000
25.000
N2
N3
N5
SECTION A-A
SHELL NOZZLES
SECTION B-B
STATIONARY HEAD NOZZLES
TRUE ORIENTATION OF NOZZLES VIEWED FROM STATIONARY HEAD END

FIG. 1.2.—*contd.*

1.6. HTFS and HTRI

An introduction to heat exchanger technology would be incomplete without mentioning the two major information services in the field. These are the Heat Transfer and Fluid Flow Service (HTFS) operated jointly by Harwell and the National Engineering Laboratory, and Heat Transfer Research Incorporated (HTRI) of Alhambra, Los Angeles. These organisations provide a similar service. In the case of HTFS 'the sponsors receive the following information in return for their annual fee:

(a) *Design Reports*
These are concerned with the design of particular types of equipment or with general subjects such as fouling, heat exchanger selection and extended surface heat transfer. The aim is to provide an authoritative and easily used source to which the designer or user may refer.

(b) *Computer Programs*
These are produced as an aid to design for those cases where manual calculation is impracticable or uneconomic.

(c) *Consultancy*
Each member has the right to consult HTFS in confidence about its problems in heat transfer, fluid flow and related topics. Two man-days of such consultancy are allowed per annum as part of the subscription service; extra may be paid for if required.

(d) *Information Office*
Sponsors have the right to make use of the HTFS information service in finding data on heat transfer and fluid flow and on related matters such as fluid physical properties'.

There are supporting research programmes.

1.7. Future Developments

In conclusion it is relevant to take another look at the future; we have already done so in the context of the design process and computers. No doubt in the future more work will be done on precise three-dimensional solutions of the complex flows in heat exchangers. At present most heat exchangers are designed on the basis of one-dimensional models, modified to allow for two- or three-dimensional effects.

Energy conservation will also presumably lead to pressure for improved heat transfer and heat exchanger systems; 9% of the articles in the bibliography, which with a few exceptions were randomly selected, concern the heat pump. Pressure will also exist for reduced noise, reduced pollution, increased safety and improved reliability. It is always difficult to project to

TABLE 1.2

FUTURE RESEARCH ACTIVITIES IN HEAT TRANSFER

Motivating force	*Activity*
Developments in computers: 3-dimensional programs	Shell-side flows: condensation boiling temperature-dependent fluids Combustion
Energy conservation: systems	Improved process cycles Network optimisation Heat pumps Drying (environments) Plant control
Energy conservation: waste heat recovery	Cheap constructions Heat pipe exchangers Gas-side fouling Expendable exchangers
Energy conservation: general	Enhanced surfaces Combustion control Heat meters Thermal storage Intermittent heating (insulation)
Miscellaneous: pollution reduction safety, reliability noise reduction	Flow instabilities Direct contact heat transfer Fluidised-bed combustion Fluidised-bed heat transfer Solar collectors Heat transfer fluids Fouling and prevention
General heat transfer problems: manufacturing domestic medical	Heat pipes

the future, but with this background Table 1.2 illustrates the likely areas of future research in heat exchanger technology.

REFERENCES

1. Rathore, R. N. S. and Powers, G. J. A forward branching scheme for the synthesis of energy recovery systems, *IEC Process Des. Dev.*, 1975, **14**(2), 175–81.

2. CHALLAND, T. B. Computer systems help boost heat recovery, *Oil Gas J.*, 1976, **74**(36), 105.
3. BOLAND, D. and LINNHOFF, B. The preliminary design of networks for heat exchange by systematic methods, *Chem. Eng.*, 1979, **57**(343), 222–28.
4. NISHIDA, N., LIU, Y. A. and LAPIDUS, L. Studies in chemical process design and synthesis. Part III: A simple and practical approach to the optimal synthesis of heat-exchanger networks, *AIChE J.*, 1977, **23**(1), 77–93.
5. CAD 74. *International Conference on Computers in Engineering and Building Design, 25–27 September 1974*, Imperial College, London, UK.
6. CHEN, C. C. and EVANS, L. B. More computer programs for chemical engineers, *Chem. Eng.*, 1979, **86**(11), 167–73.
7. ANON. The future of real-time technology: A report to the Computers, Systems and Electronics Requirements Board, August 1975, Department of Industry, London.
8. MURRAY, I. Speed up exchanger design with STEM, *Heat. Vent. Eng.*, February 1977, **51**(595), 5–8.
9. ANON. Heat exchanger design service saves time, *Processing*, 1978, **24**(6), 53–55.
10. CUNNINGHAM, C. Software systems cut design fatigue, *Processing*, June 1979, **25**(6), 25–27.
11. GIBBS, P. J. Computer-aided integrated systems, *Chart. Mech. Eng.*, February 1980, **27**(2), 84–88.
12. SOPER, B. M. H. Flow-induced vibration in shell-and-tube heat exchangers, in: *Proceedings of International Meeting: Industrial Heat Exchangers and Heat Recovery, 14–16 November 1979*, Association of Engineers, Graduate University of Liege (A.I.Lg), Paper B, 1–18.

Chapter 2

SHELL-AND-TUBE EXCHANGERS FOR SINGLE-PHASE APPLICATION

I. D. R. GRANT

National Engineering Laboratory, East Kilbride, Glasgow, UK

SUMMARY

Reference is made early on to the different types of shell-and-tube heat exchanger used in chemical and process plant. This is followed by methods of calculating the temperature driving force for heat transfer for the different types. The conditions which have to be satisfied for these methods to be valid are given in some detail. The calculation of tube-side pressure drop and heat transfer is also dealt with. The opportunity is taken here to describe recent correlations by Churchill for friction factor and heat transfer coefficient which cover the full Reynolds number range. In dealing with shell-side pressure drop and heat transfer, two methods are described. The first by Bell is considered more suitable for hand calculation while the second, an adaptation of the Tinker Method, is more suited for computer use. The second is now extensively used in shell-and-tube computer programs and in particular in the programs produced by the Heat Transfer and Fluid Flow Service (HTFS). The chapter finishes with the description of a new type of baffle which is gaining popularity since it overcomes the problem of heat exchanger failure due to excessive tube vibration.

NOTATION

A	Tube outside surface area (m^2)
A'	Parameter defined by eqn. (14)
a	Coefficient in eqn. (51)

B	Coefficient in Blasius equation
B'	Parameter defined by eqn. (15)
C	Contraction coefficient
C_p	Specific heat (J/kg K)
D	Characteristic length (m)
D_e	Equivalent diameter (m)
D_I	Tube inside diameter (m)
D_O	Tube outside diameter (m)
D_v	Volumetric equivalent diameter (m)
F	Correction factor to LMTD
F'	Flow fraction
F_b	Correction factor for by-passing
F_l	Correction factor for leakage
F_n	Correction factor for number of rows
F_w	Correction factor for window geometry
f	Fanning friction factor
f'	Friction factor defined in eqn. (31)
f''	Friction factor defined in eqn. (16)
G	Mass velocity (kg/m^2 s)
G_z	Geometric mean mass velocity (kg/m^2 s)
G_1	Inlet mass velocity (kg/m^2 s)
G_2	Outlet mass velocity (kg/m^2 s)
h	Heat transfer coefficient (W/m^2 K)
h_I	Shell-side heat transfer coefficient for ideal tube bundle (W/m^2 K)
h_{io}	Tube-side heat transfer coefficient referred to shell side (W/m^2 K)
h_o	Overall shell-side heat transfer coefficient (W/m^2 K)
K	Parameter defined in eqn. (38)
K'	Number of velocity heads
K''	Pressure loss coefficient
K''_R	Pressure loss coefficient at reference condition
k	Thermal conductivity (W/m K)
L	Tube length (m)
M	Shell-side total mass flow (kg/s)
m	Exponent in eqn. (51)
N	Number of tube rows
N_B	Number of baffles
N_{CF}	Effective number of crossflow restrictions
Nu	Nusselt number
Nu_l	Nusselt number defined in Table 2.1
Nu_{lc}	Nusselt number defined in Table 2.1

Nu_0^0 Nusselt number defined in Table 2.1
Nu_R Nusselt number at reference condition
N_W Number of restrictions for crossflow in window
n Exponent in Blasius type equation
P Thermal effectiveness
Pr Prandtl number
ΔP Pressure drop (N/m^2)
ΔP_B Pressure drop in ideal tube bundle (N/m^2)
ΔP_BP Crossflow pressure drop corrected for by-passing (N/m^2)
ΔP_L Shell-side pressure drop with leakage (N/m^2)
ΔP_NL Shell-side pressure drop without leakage (N/m^2)
ΔP_Ni Pressure drop at inlet nozzle (N/m^2)
ΔP_No Pressure drop at outlet nozzle (N/m^2)
ΔP_R Pressure drop in tube bundle headers (N/m^2)
ΔP_TOT Overall shell-side pressure drop (N/m^2)
ΔP_W Pressure drop in baffle window (N/m^2)
Q Heat transfer rate per unit area (W)
R Heat capacity
R_D Fouling resistance (m^2 K/W)
Re Reynolds number
S Cross-sectional area (m^2)
S_l Longitudinal tube pitch (m)
S_t Transverse tube pitch (m)
S_1 Inlet cross-sectional area (m^2)
S_2 Outlet cross-sectional area (m^2)
s Tube pitch (m)
T_1 Shell-side inlet temperature (K)
T_2 Shell-side outlet temperature (K)
ΔT_M Mean temperature difference (K)
ΔT_LM Logarithmic mean temperature difference (K)
t_1 Tube-side inlet temperature (K)
t_2 Tube-side outlet temperature (K)
U Overall heat transfer coefficient (W/m^2 K)
U_1 Overall heat transfer coefficient at inlet (W/m^2 K)
U_2 Overall heat transfer coefficient at outlet (W/m^2 K)
X_l Longitudinal tube pitch-to-diameter ratio
X_t Transverse tube pitch-to-diameter ratio
β Correction factors in eqns. (25) and (50)
ε Effective roughness (m)
μ Fluid dynamic viscosity (N s/m^2)

ρ Fluid density (kg/m^3)
ρ_1 Inlet fluid density (kg/m^3)
ρ_2 Outlet fluid density (kg/m^3)

Subscripts
A Tube/baffle leakage flow
B Crossflow between baffle
b At bulk fluid conditions
bm At mean bulk fluid conditions
C Bundle shell by-pass flow
E Baffle/shell leakage flow
F Pass-partition flow
WB Baffle window flow through bundle
WBP Baffle window flow by-passing bundle
W At wall conditions

1. INTRODUCTION

The thermal design of shell-and-tube heat exchangers for single-phase flow is primarily concerned with the calculation of overall heat transfer coefficient, temperature driving force and flow pressure drops.

The rate of heat transfer per unit surface area is given by the product of an overall heat transfer coefficient and a mean temperature difference between the tube-side and shell-side streams

$$\frac{Q}{A} = U \Delta T_M \tag{1}$$

where,

U is the overall heat transfer coefficient based on the tube outside surface area (W/m^2 K)
A is the tube outside surface area between tubesheets, and
ΔT_M is the mean temperature difference between tube-side and shell-side streams (K).

The overall heat transfer coefficient U or the overall resistance $1/U$ is equal to the sum of the individual resistances to heat transfer between tube-side and shell-side fluids

$$\frac{1}{U} = \frac{1}{h_{io}} + \frac{1}{h_o} + \frac{D_O \ln \frac{D_O}{D_I}}{2k} + R_D \tag{2}$$

where,

h_{io} is the tube-side coefficient referred to the tube outside area (W/m^2 K),
h_o is the shell-side coefficient (W/m^2 K),
D_I and D_O are the tube inside and outside diameters (m),
k is the tube thermal conductivity (W/m K), and
R_D is the fouling resistance referred to the tube outside area (m^2 K/W).

2. TYPES OF HEAT EXCHANGER

A list of different types of heat exchanger is given in TEMA.[1] A system of designating a letter to the different shells, and front and rear heads of the shell allows a heat exchanger to be designated, for example, a TEMA type AES. This refers to a split-ring floating head heat exchanger with removable cover and a single-pass shell. This is explained further with the assistance of Fig. 2.1.

A TEMA type AES is shown in greater detail in Fig. 2.2. Numbers are used in Fig. 2.2 to identify the parts according to the list therein.

3. MEAN TEMPERATURE DIFFERENCE

The mean temperature difference ΔT_M or temperature driving force for heat transfer is usually obtained from the product of a logarithmic mean temperature difference (LMTD) and a correction factor F

$$\Delta T_M = \Delta T_{LM} F \tag{3}$$

where ΔT_{LM} is the logarithmic mean of the inlet and outlet temperature differences at each end of the heat exchanger.

3.1. E-type Shell

The mean temperature difference for a single-pass unit as shown in Fig. 2.3(a) is given by the logarithmic mean temperature difference, providing the following conditions apply:

(a) flow rate of each fluid is constant,
(b) specific enthalpy of each fluid is linear with temperature,
(c) overall heat transfer coefficient is constant throughout the unit, and
(d) heat losses are negligible.

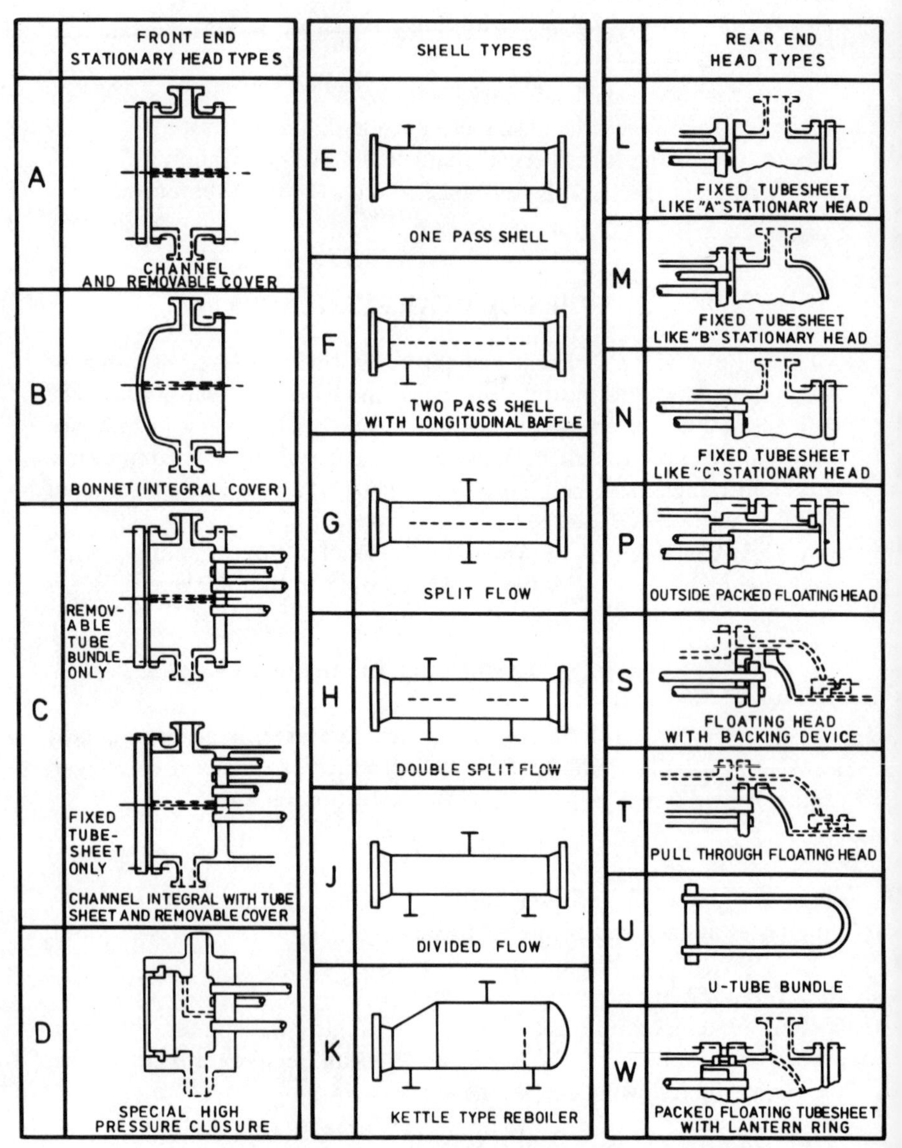

FIG. 2.1. TEMA nomenclature of heat exchanger types.[1] ©—1978 by Tubular Exchanger Manufacturers' Association.

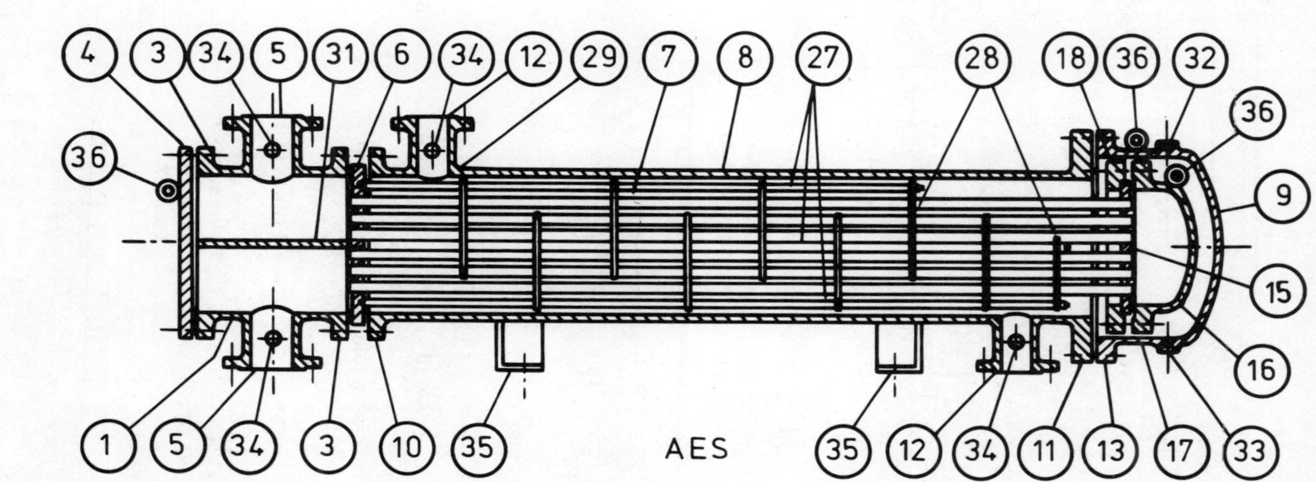

FIG. 2.2. TEMA type AES.[1] Heat exchanger part terminology: 1, stationary head—channel; 3, stationary head flange—channel or bonnet; 4, channel cover; 5, stationary head nozzle; 6, stationary tubesheet; 7, tubes; 8, shell; 9, shell cover; 10, shell flange—stationary head end; 11, shell flange—rear head end; 12, shell nozzle; 13, shell cover flange; 15, floating tubesheet; 16, floating head cover; 17, floating head flange; 18, floating head backing device; 27, tie rods and spacers; 28, transverse baffles or support plates; 29, impingement baffle; 31, pass partition; 32, vent connection; 33, drain connection; 34, instrument connection; 35, support saddle; 36, lifting lug. ©—1978 by Tubular Exchanger Manufacturers' Association.

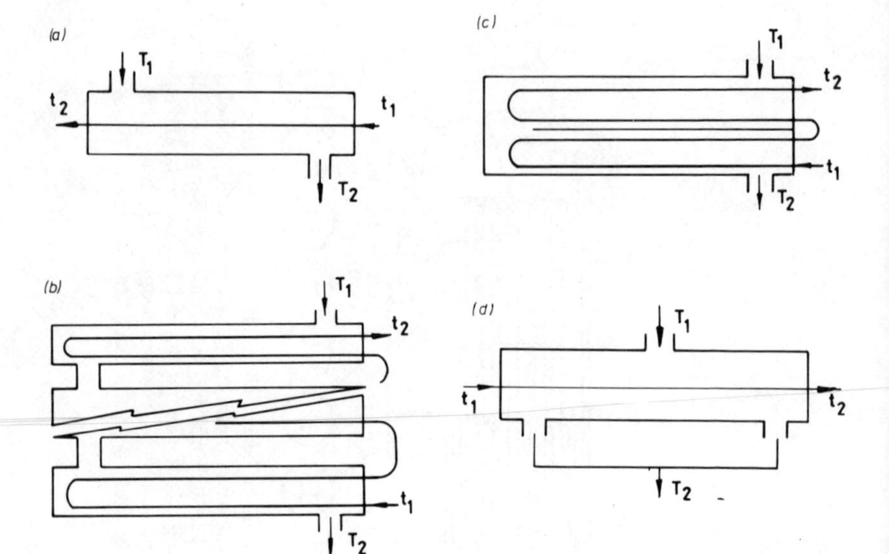

FIG. 2.3. Shell flow arrangements. (a) E-type shell, (b) E-type shells connected in series, (c) F-type shell, (d) J-type shell.

The LMTD for countercurrent flow is

$$\Delta T_{\mathrm{LM}} = \frac{(T_1 - t_2) - (T_2 - t_1)}{\ln\left(\dfrac{T_1 - t_2}{T_2 - t_1}\right)} \tag{4}$$

where,

T_1 and T_2 are the shell-side inlet and outlet temperatures respectively (K), and

t_1 and t_2 are the tube-side inlet and outlet temperatures respectively (K).

For concurrent flow the terminal temperature differences are $(T_1 - t_1)$ and $(T_2 - t_2)$.

If the overall heat transfer coefficient varies linearly with temperature in a single-pass unit, eqn. (4) can be written in the form

$$Q = \frac{A\{U_2(T_1 - t_2) - U_1(T_2 - t_1)\}}{\ln\left\{\dfrac{U_2(T_1 - t_2)}{U_1(T_2 - t_1)}\right\}} \tag{5}$$

If $U_1 = U_2$ eqn. (5) reduces to eqn. (4) which is the LMTD for countercurrent flow.

If neither (c) nor (d) is satisfied then Emerson[2] and Butterworth[3] should be consulted.

3.2. E-type Shells in Series

In multi-pass heat exchangers as shown in Fig. 2.3(b) the flow is both countercurrent and concurrent. The mean temperature difference then lies between the temperature difference of eqn. (4) for countercurrent flow and a similar equation with terminal temperature differences $(T_1 - t_1)$ and $(T_2 - t_2)$ for concurrent flow.

The mean temperature difference is therefore calculated using eqn. (4) with the appropriate value of correction factor F. Two further conditions in addition to those in Section 3.1 now apply. These are:

- (e) equal heat transfer surface in each pass, and
- (f) temperature of shell-side fluid in any pass uniform over cross-section (i.e. complete mixing).

The correction factor F for one shell pass and two or more even numbers of tube passes is given by

$$F = \frac{\sqrt{(R^2+1)}\ln\left(\dfrac{1-P}{1-RP}\right)}{(R-1)\ln\left[\dfrac{2-P\{R+1-\sqrt{(R^2+1)}\}}{2-P\{R+1+\sqrt{(R^2+1)}\}}\right]} \tag{6}$$

where the heat capacity R is given by

$$R = \frac{T_1 - T_2}{t_2 - t_1} \tag{7}$$

and the thermal effectiveness by

$$P = \frac{t_2 - t_1}{T_1 - t_1} \tag{8}$$

Equations are available for the calculation of F for two and three or more shell passes with an even number of tube passes. The F values are also readily obtained from graphs[1] as shown in Fig. 2.4. Similar graphs also give F values for shells with an odd number of tube passes.

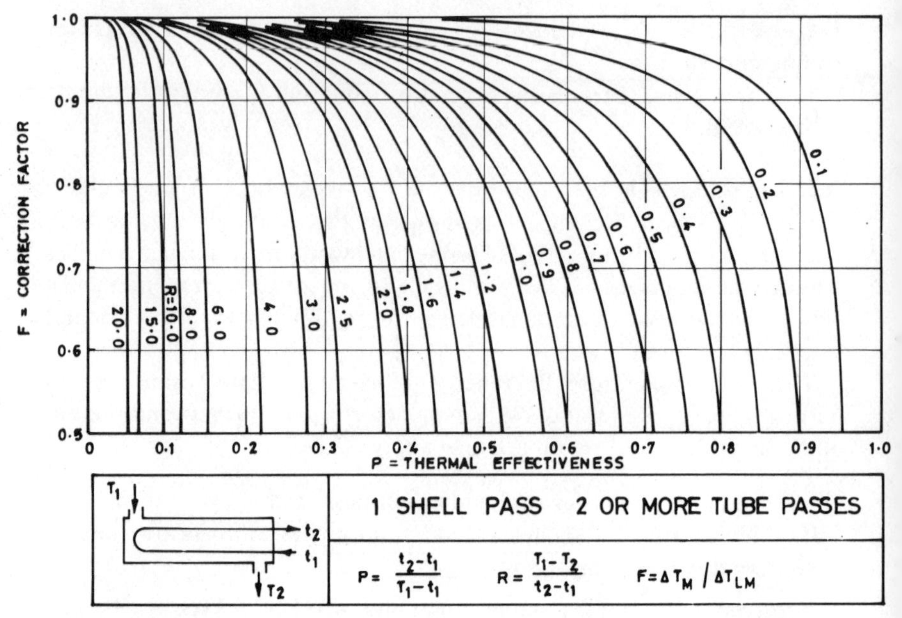

FIG. 2.4. Correction factor F.[1] ©—1978 by Tubular Exchanger Manufacturers' Association.

3.3. F-type Shell

A further condition must be added to conditions (a)–(f) for an F-shell. This is:

(g) no fluid or heat leakage must occur between shell passes.

The mean temperature difference for an F-shell illustrated in Fig. 2.3(c) is given by

$$F = \frac{\sqrt{(R^2+1)}\ln\left(\dfrac{1-P}{1-RP}\right)}{2(R-1)\ln\left[\dfrac{\dfrac{2}{P}-1-R+\dfrac{2}{P}\sqrt{\{1-P)(1-RP)\}}+\sqrt{(R^2+1)}}{\dfrac{2}{P}-1-R+\dfrac{2}{P}\sqrt{\{(1-P)(1-RP)\}}-\sqrt{(R^2+1)}}\right]} \qquad (9)$$

An F-shell is equivalent to two E-shells connected in series (four tube passes). The convection factor for F-shells connected in series is therefore obtained from an equivalent arrangement of E-shells.

3.4. J-type Shell

For a J-shell as shown in Fig. 2.3(d), condition (g) is modified to:

(g) no fluid or heat leakage between flows in different halves of shell.

and a further condition added:

(h) equal flow rate of shell-side fluid in each half.

The F values for a J-shell with one, two, four or more tube passes are given in a paper by Jaw.[4]

4. TUBE-SIDE PRESSURE DROP AND HEAT TRANSFER

The pressure drop for flow inside a straight tube is given by the Fanning equation

$$\Delta P = \frac{4fLG^2}{2\rho_{bm}D_e} \tag{10}$$

where,

f is the friction factor,
L is the tube length (m),
G is the fluid mass velocity (kg/m^2 s),
ρ_{bm} is the fluid density at bulk mean temperature (kg/m^3), and
D_e is the equivalent diameter which is the tube inside diameter (m).

Friction factors are required for the different flow regimes which are described as laminar, transitional or turbulent flow. These regimes are determined by the following criteria

$\text{Re} < 2000$ for laminar flow

$\text{Re} \geq 2000$ for transitional or turbulent flow

where the Reynolds number Re is defined by

$$\text{Re} = \frac{GD_e}{\mu} \tag{11}$$

where μ is the fluid dynamic viscosity (N s/m^2).

The flow at the ends of the tubes including entrance, exit and return losses is difficult to predict accurately and it is common practice to assume that a number of velocity heads are lost, that is

$$\Delta P_{R} = \frac{4K'G^2}{2\rho} \tag{12}$$

where K' is the number of velocity heads.

4.1. Friction Factor

A single equation has been developed by Churchill[5] to correlate tube friction factor over the full range of Reynolds numbers. The equation (strictly for incompressible isothermal flow) is given by

$$f'' = \left\{\left(\frac{8}{\mathrm{Re}}\right)^{12} + \frac{1}{(A' + B')^{3/2}}\right\}^{1/12} \tag{13}$$

where

$$A' = \left\{2{\cdot}457 \ln \frac{1}{\left(\frac{7}{\mathrm{Re}}\right)^{0{\cdot}9} + \frac{0{\cdot}27\varepsilon}{D_e}}\right\}^{16} \tag{14}$$

and

$$B' = \left(\frac{37\,530}{\mathrm{Re}}\right)^{16} \tag{15}$$

The Reynolds number is given by eqn. (11) and ε is the tube effective roughness. Values for ε/D_e are tabulated in several different books including Perry's *Chemical Engineers Handbook*.

This friction factor is related to the more commonly used Fanning friction factor of eqn. (10) by

$$f = 2f'' \tag{16}$$

A plot of friction factor f against Reynolds number for the laminar, transition and fully developed turbulent flow regimes for $\varepsilon/D_I = 0{\cdot}000\,001$ is shown in Fig. 2.5. This curve corresponds to flow in a smooth tube. Commercial tubes may have higher friction factors in the fully turbulent regime. A feature of eqn. (13) as shown in Fig. 2.5 is that it gives unique values in the transition regime, although this does not remove the uncertainty in this regime.

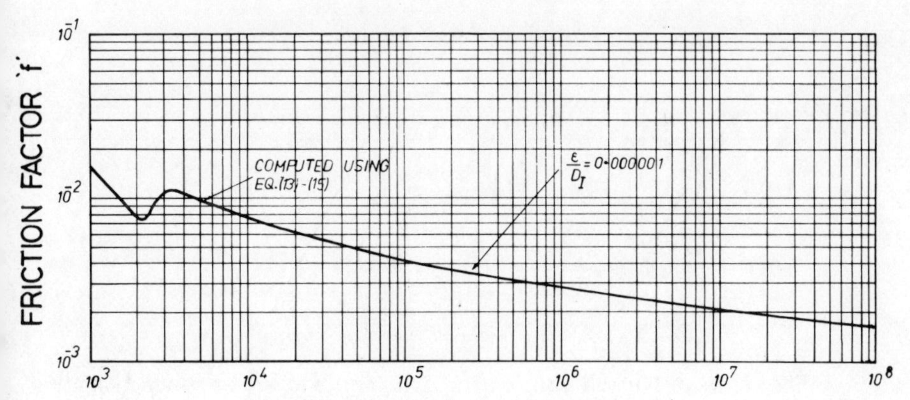

FIG. 2.5. Friction factor for smooth tube.

4.2. Heat Transfer Coefficient

A Nusselt number Nu is defined by

$$\mathrm{Nu} = \frac{hD_{\mathrm{I}}}{k} \tag{17}$$

where,

h is the heat transfer coefficient (kW/m^2 K),
D_{I} is the tube inside diameter (m), and
k is the fluid thermal conductivity (W/m K).

Churchill[6] also gives a single correlating equation for fully developed turbulent flow in smooth tubes. The equation which covers all Reynolds and Prandtl numbers is

$$(\mathrm{Nu})^{10} = (\mathrm{Nu}_{\mathrm{l}})^{10} + \left[\frac{\exp(2200-\mathrm{Re})/365}{(\mathrm{Nu}_{\mathrm{lc}})^2} + \left\{\frac{1}{\mathrm{Nu}_0^0 + \dfrac{0{\cdot}079\mathrm{Re}\sqrt{(f''\mathrm{Pr})}}{(1+\mathrm{Pr}^{4/5})^{5/6}}}\right\}^2\right]^{-5} \tag{18}$$

Values of Nu_{l}, $\mathrm{Nu}_{\mathrm{lc}}$ and Nu_0^0 have to be substituted in the equation for either *uniform heat flux* (UHF) or *uniform wall temperature* (UWT) conditions, whichever applies. The equation is strictly for near-isothermal conditions.

Prandtl number is defined by

$$\mathrm{Pr} = \frac{C_{\mathrm{p}}\mu}{k} \tag{19}$$

where C_{p} is the specific heat (J/kg K).

TABLE 2.1
NUSSELT NUMBER VALUES

Conditions	Nu_l	Nu_{lc}	Nu_0^0
UWT	$3{\cdot}657\left\{1+\left(\frac{\text{RePr}D_\text{I}}{7{\cdot}6L}\right)^{8/3}\right\}^{1/8}$	$3{\cdot}657\left\{1+\left(\frac{276\text{Pr}D_\text{I}}{L}\right)^{8/3}\right\}^{1/8}$	4·8
UHF	$4{\cdot}364\left\{1+\left(\frac{\text{RePr}D_\text{I}}{7{\cdot}3L}\right)^{2}\right\}^{1/6}$	$4{\cdot}364\left\{1+\left(\frac{287\text{Pr}D_\text{I}}{L}\right)^{2}\right\}^{1/6}$	6·3

The values of Nusselt number Nu_l, Nu_lc and Nu_0^0 are obtained using the methods shown in Table 2.1. Equation (13) in Section 4.1 can be used for friction factor f''.

5. SHELL-SIDE PRESSURE DROP AND HEAT TRANSFER

Flow on the shell side is complex due to manufacturing clearances and tolerances. This is best illustrated by the flow model of Tinker,[7] Fig. 2.6, which shows that in addition to flow across the bundle (crossflow) and flow around the segmental baffle (window flow) there is bundle by-pass flow and tube/baffle and baffle/shell leakage flow. The Tinker model is now the basis of methods used in present day computer programs for solving the

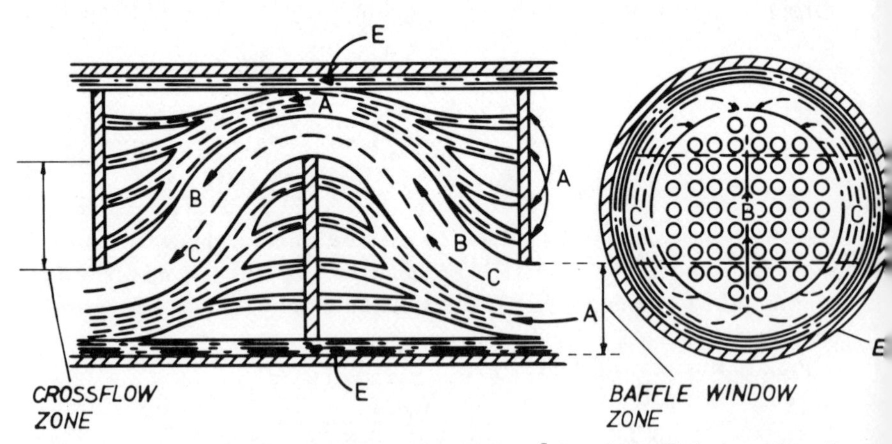

FIG. 2.6. Shell-side flow streams. (After Tinker.)[7] A, tube/baffle leakage flow; B, crossflow between baffles; C, bundle/shell by-pass flow; E, baffle/shell leakage flow.

distribution of the total flow through these different areas and hence solving the pressure drop. An example of how this flow model is used is given later in Section 5.4.

The ideal crossflow pressure drop (for rectangular tube bank with half-tubes on walls) is given by

$$\Delta P_{B} = \frac{K'' N_{CF} G_{max}^{2}}{2\rho_{bm}} \tag{20}$$

where,

K'' is the overall pressure loss coefficient,
N_{CF} is the effective number of crossflow restrictions,
G_{max} is the mass velocity at the minimum flow area (kg/m^2 s), and
ρ_{bm} is the fluid density at bulk mean temperature (kg/m^3).

For turbulent flow the window pressure drop is given by

$$\Delta P_{W} = \frac{(2 + 0{\cdot}6N_{W})G_{z}^{2}}{2\rho_{bm}} \tag{21}$$

where,

G_z is the geometric mean mass velocity (kg/m^2 s),
N_W is the number of restrictions for crossflow in the window, and
ρ_{bm} is the fluid density at bulk mean temperature (kg/m^3).

The pressure drop for the shell nozzles, i.e. the inlet and outlet nozzles may be taken as equivalent to a given number of velocity heads based on the velocity in the nozzle (see eqn. (12)). Another way is to treat the inlet nozzle as a sudden expansion such that

$$\Delta P_{Ni} = \frac{G_1^2 S_1}{\rho_1 S_2}\left(\frac{S_1}{S_2} - 1\right) \tag{22}$$

where,

S_1 is the nozzle area (m^2),
S_2 is the bundle entry area (m^2),
G_1 is the mass velocity based on the nozzle area (kg/m^2 s), and
ρ_1 is the fluid density at inlet temperature (kg/m^3).

Area A_2 is not so easily defined and will depend upon whether an impingement baffle is fitted or not. If such a baffle is fitted, an additional velocity head will be lost due to the 90° turn around the baffle.

The outlet nozzle may be treated as a sudden contraction, thus

$$\Delta P_{\mathrm{No}} = \frac{G_2^2}{2\rho_2}\left\{1 - \left(\frac{S_2}{S_1}\right)^2 + \left(\frac{1}{C} - 1\right)^2\right\} \tag{23}$$

where,

S_1 is the bundle exit area (m^2),
S_2 is the nozzle area (m^2),
G_2 is the mass velocity based on the nozzle area ($kg/m^2\,s$), and
C is the contraction coefficient ($C = \frac{2}{3}$).

5.1. Tube Bank Parameters

Before we can use eqn. (20) to calculate the crossflow pressure drop, a number of tube bank parameters have to be defined and evaluated.

Tube Pitch to Diameter Ratio

The value of the transverse tube pitch to diameter ratio $X_t = S_t/D_O$ and the longitudinal tube pitch to diameter ratio $X_l = S_l/D_O$ depends upon tube arrangement. This is illustrated in Table 2.2 which is taken from an ESDU publication.[8]

TABLE 2.2
VALUES OF TUBE PITCH-TO-DIAMETER RATIO

Tube arrangement	*Bank type*	*To obtain* X_t *multiply* s/D *by*	*To obtain* X_l *multiply by*
Flow → Square in-line	In-line	1·0	1·0
Flow → Rotated square	Staggered	1·414	0·707
Flow → Equilateral triangle	Staggered	1·0	0·866
Flow → Rotated equilateral triangle	Staggered	0·866	0·5

Maximum Mass Velocity

The maximum mass velocity for crossflow is defined by

$$G_{max} = \frac{M}{S_{min}} \tag{24}$$

where S_{min} is the minimum flow area normal to the flow. This is usually calculated along a straight path although for a staggered tube arrangement a staggered path may give the minimum area. This is illustrated in Fig. 2.7 which is again taken from ESDU.

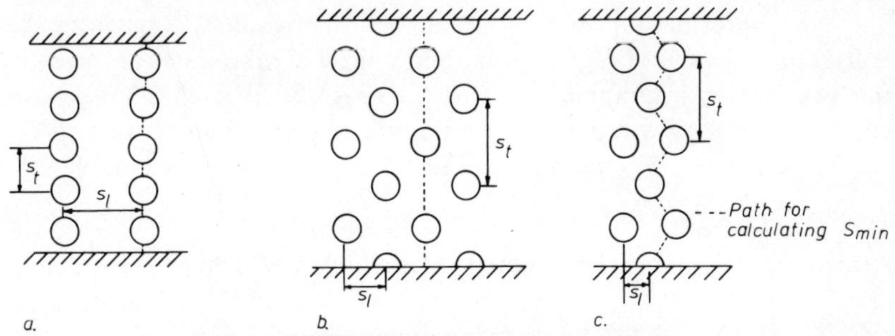

FIG. 2.7. Path for calculating minimum flow area. (After ESDU.)[8] (a), In-line bank; (b), staggered bank; (c), staggered bank.

Number of Tube Rows

In eqn. (20) N_{CF} is the number of restrictions through the tube bank. For in-line and staggered tube banks where S_{min} is calculated in a transverse plane (see Figs. 2.7(a) and (b)) $N_{CF} = N$ the number of rows, i.e. two in Fig. 2.7(a) and four in Fig. 2.7(b). For staggered banks where S_{min} is calculated along a staggered path as in Fig. 2.7(c), $N_{CF} = N - 1 = 2$.

5.2. Pressure Loss Coefficient

The overall pressure loss coefficient for a tube bank depends upon the effective number of restrictions N_{CF} in the flow direction. Methods of estimating this coefficient for one row, 10 or more rows and between two and nine rows inclusive are given in the ESDU publication.[8] Here it will be sufficient to give the method for 10 or more rows, which is the most likely requirement for a shell-and-tube heat exchanger.

The value of overall loss coefficient K for 10 or more rows is given by[8]

$$K'' = K''_R \beta_1 \beta_2 \beta_3 \beta_4 \tag{25}$$

TABLE 2.3

Factor	*Description*
β_1	Allows for variation of fluid properties due to $T_w \neq T_b$
β_2	Allows for by-passing between tubes and duct walls
β_3	Allows for tube inclination to crossflow
β_4	Allows for tube roughness ratios different from $\varepsilon/D = 5 \times 10^{-5}$

where K_R'' is the loss coefficient at a reference condition. Factors β_1 to β_4 are used to correct K_R'' for the effects shown in Table 2.3.

Further information on how to calculate the β-values is given in the ESDU item. Values of K_R'' for in-line and staggered tube banks are given in this publication in graphical form for a range of X_l and X_t values. An equilateral triangle arrangement with a tube pitch to diameter ratio S/D_O (see Table 2.2) of 1·5 giving $X_t = 1{\cdot}5$ and $X_l = 1{\cdot}3$, is illustrated in Fig. 2.8.

5.3. Bell Method

A method which avoids the flow model analysis of Tinker[7] in order to

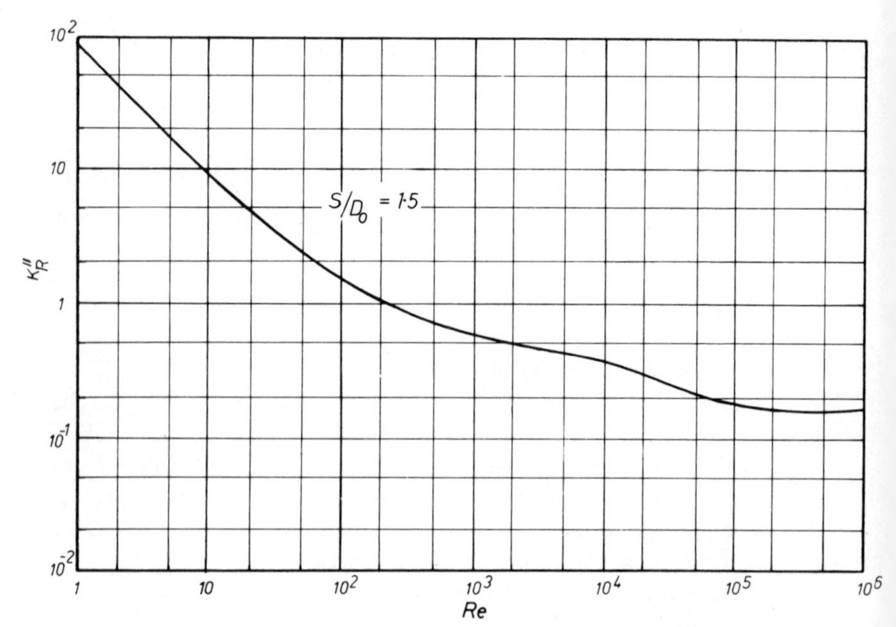

FIG. 2.8. Pressure loss coefficient for staggered arrangements.

calculate the shell-side pressure drop of segmentally baffled shell-and-tube heat exchangers is that of Bell.[9] The basis of this method is to consider a segmentally baffled heat exchanger as a series of ideal tube banks connected by window zones. The flow through this series is short circuited due to bundle by-passing and leakage across the baffles. The overall shell-side pressure drop less the nozzle pressure drop is given by

$$\Delta P_{\mathrm{TOT}} = 2\,\Delta P_{\mathrm{BP}}\left(1 + \frac{N_{\mathrm{W}}}{N_{\mathrm{CF}}}\right) + \{(N_{\mathrm{B}} - 1)\,\Delta P_{\mathrm{B}} + N_{\mathrm{B}}\,\Delta P_{\mathrm{W}}\}\left(\frac{\Delta P_{\mathrm{L}}}{\Delta P_{\mathrm{NL}}}\right) \quad (26)$$

where,

ΔP_{BP} is the pressure drop in a crossflow zone corrected for by-passing (N/m^2),
ΔP_{L} is the overall pressure drop for the exchanger with leakage (N/m^2),
ΔP_{NL} is the overall pressure drop for no leakage (N/m^2),
ΔP_{W} is the pressure drop for a window zone (N/m^2),
N_{CF} is the number of restrictions in the crossflow zone, i.e. the number of rows between baffle tips for square and equilateral triangle arrays, and
N_{W} is the number of restrictions for crossflow in the window.

The first term on the right-hand side shows that the two end zones are treated as non-leakage parts. Another way of writing eqn. (26) is

$$\Delta P_{\mathrm{TOT}} = 2\,\Delta P_{\mathrm{B}} F_{\mathrm{b}}\left(1 + \frac{N_{\mathrm{W}}}{N_{\mathrm{CF}}}\right) + \{(N_{\mathrm{B}} - 1)\,\Delta P_{\mathrm{B}} F_{\mathrm{b}} + N_{\mathrm{B}}\,\Delta P_{\mathrm{W}}\} F_{\mathrm{l}} \quad (27)$$

where,

F_{b} is the by-pass correction factor,
F_{l} is the leakage correction factor, and
ΔP_{B} is the pressure drop for an ideal tube bank, i.e. no by-pass or leakage.

5.4. Tinker Method

The flow model of Tinker[7] for a segmentally baffled shell-and-tube heat exchanger is shown in Fig. 2.6. The different shell-side flows are labelled after the manner of Tinker as follows:

A tube/baffle leakage flow,
B crossflow between baffles,
C bundle/shell by-pass flow, and
E baffle/shell leakage flow.

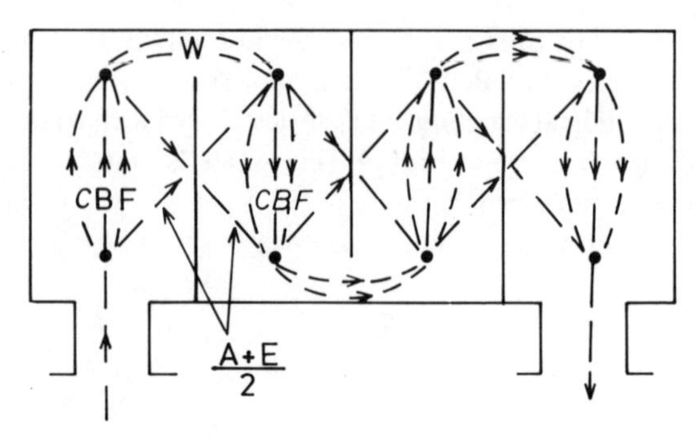

FIG. 2.9. Flow model used by Grant.[10,11] A, Tube/baffle leakage flow; B, crossflow; C, by-pass flow; E, baffle/shell leakage flow; F, pass–partition flow; W, window bundle and by-pass flow.

In order to set up equations to calculate the rates of these different flows Grant[10,11] uses the flow model of Fig. 2.9. In this model two nodal points are selected in each baffle space, level with the baffle-cut edges. The lines connecting the nodes represent the different flows. The additional flows in Fig. 2.9 when compared with the Tinker model of Fig. 2.6 are

F pass-partition flow,
W_{B} baffle window flow through bundle, and
W_{BP} baffle window flow by-passing bundle.

5.4.1. Nodal Mass Balance

Let F_{i}' denote the fraction of the total flow M in any flow path. A mass balance is carried out at each node. In the middle region of the heat exchanger, that is between the first and last baffles, it is only necessary because of symmetry to consider one crossflow zone.

A mass balance at either of the nodes in a middle crossflow zone gives

$$F_{\mathrm{WB}}' + F_{\mathrm{WBP}}' - F_{\mathrm{B}}' - F_{\mathrm{C}}' - F_{\mathrm{F}}' = 0 \qquad (28)$$

When the total flow along the heat exchanger is considered

$$F_{\mathrm{A}}' + F_{\mathrm{E}}' + F_{\mathrm{WB}}' + F_{\mathrm{WBP}}' - 1 = 0 \qquad (29)$$

A mass balance at the first node in the inlet region of the heat exchanger gives

$$\frac{F_{\mathrm{A}}'}{2} + F_{\mathrm{B}}' + F_{\mathrm{C}}' + \frac{F_{\mathrm{E}}'}{2} + F_{\mathrm{F}}' - 1 = 0 \qquad (30)$$

A mass balance at the second node combined with eqn. (29) yields eqn. (30) again.

A mass balance at both nodes in the outlet region of the heat exchanger also yields eqn. (30).

The flow in the middle region of a segmentally baffled shell-and-tube heat exchanger according to the flow model of Fig. 2.9 is given by eqns. (28) and (29) and in either end region by eqns. (29) and (30).

5.4.2. *Flow Fractions*

The flow fraction in any path is defined by the Fanning equation (used previously in Section 4)

$$F_i' = \left(\frac{2\,\Delta P_i \rho S_i^2}{f_i' M^2}\right)^{1/2} \tag{31}$$

where,

ΔP_i is the pressure drop along the path,
S_i is the cross-sectional area for flow,
f_i' is the friction factor, and
M is the total shell-side flow.

The crossflow, by-pass, and pass-partition friction factors become

$$f_i' = 4 f_i N \tag{32}$$

where N is the effective number of tube rows, that is the number of major restrictions to flow. The tube-baffle and baffle-shell leakage friction factor is simply

$$f_i' = f_i \tag{33}$$

Let the friction factor f_i for any path be given by a Blasius type equation

$$f_i = \frac{B_i}{\mathrm{Re}_i^{ni}} \tag{34}$$

where,

$$\mathrm{Re}_i = \frac{F_i' M D_i}{S_i \mu} \tag{35}$$

The pressure drops for crossflow, by-pass, and pass-partition flow are equal as shown in Fig. 2.9 and for convenience this pressure drop is denoted by ΔP_{CF}.

$$\Delta P_{CF} = \Delta P_B = \Delta P_C = \Delta P_F \tag{36}$$

The pressure drops for tube-baffle and baffle-shell leakage flow are also equal. This pressure drop is the sum of the crossflow and window zone pressure drops.

$$(\Delta P_{CF} + \Delta P_W) = \Delta P_A = \Delta P_E \tag{37}$$

Equations (31)–(37) are used to give an expression for the flow fraction in each path, for example the 'A' path fraction is

$$F_A = \{K_A(\Delta P_{CF} + \Delta P_W)\}^{\frac{1}{2-n_A}} \tag{38}$$

where,

$$K_A = \frac{2\rho S_A^{(2-n_A)} D_V^{n_A}}{B_A M^{(2-n_A)} \mu^{n_A}}$$

The flow fractions for all paths are summarised in Table 2.4.

TABLE 2.4
FLOW FRACTION EQUATIONS

Flow path	*Flow fraction*	*Coefficient* K	*Equation no.*
A	$F_A = \{K_A(\Delta P_{CF} + \Delta P_W)\}^{\frac{1}{2-n_A}}$	$K_A = \dfrac{2\rho S_A^{(2-n_A)} D_V^{n_A}}{B_A M^{(2-n_A)} \mu^{n_A}}$	(39)
B	$F_B = (K_B \Delta P_{CF})^{\frac{1}{2-n_B}}$	$K_B = \dfrac{S_B^{(2-n_B)} D_O^{n_B}}{2N_B B_B M^{(2-n_B)} \mu^{n_B}}$	(40)
C	$F_C = (K_C \Delta P_{CF})^{\frac{1}{2-n_C}}$	$K_C = \dfrac{\rho S_C^{(2-n_C)} D_O^{n_C}}{2N_C B_C M^{(2-n_C)} \mu^{n_C}}$	(41)
E	$F_E = \{K_E(\Delta P_{CF} + \Delta P_W)\}^{\frac{1}{2-n_E}}$	$K_E = \dfrac{2\rho S_E^{(2-n_E)} D_V^{n_E}}{B_E M^{(2-n_E)} \mu^{n_E}}$	(42)
F	$F_F = (K_F \Delta P_{CF})^{\frac{1}{2-n_F}}$	$K_F = \dfrac{\rho S_F^{(2-n_F)} D_O^{n_F}}{2N_F B_F M^{(2-n_F)} \mu^{n_F}}$	(43)
W_B	$F_{WB} = (K_{WB} \Delta P_W)^{\frac{1}{2-n_{WB}}}$	$K_{WB} = \dfrac{2\rho S_{WB}^{(2-n_{WB})} D_V^{n_{WB}}}{B_{WB} M^{(2-n_W)} \mu^{n_{WB}}}$	(44)
W_{BP}	$F_{WBP} = (K_{WBP} \Delta P_W)^{\frac{1}{2-n_{WBP}}}$	$K_{WBP} = \dfrac{2\rho S_{WBP}^{(2-n_{WBP})} D_V^{n_{WBP}}}{B_{WBP} M^{(2-n_{WBP})} \mu^{n_{WBP}}}$	(45)

5.4.3. Mass Balance Solution

The nodal mass balance for a middle region of the heat exchanger is obtained by the substitution of eqns. (40), (41), (43)–(45) (see Table 2.4) in eqn. (28), thus

$$(K_{WB}\,\Delta P_W)^{\frac{1}{2-n_{WB}}} + (K_{WBP}\,\Delta P_W)^{\frac{1}{2-n_{WBP}}} - (K_B\,\Delta P_{CF})^{\frac{1}{2-n_B}} - (K_C\,\Delta P_{CF})^{\frac{1}{2-n_C}} - (K_F\,\Delta_{CF})^{\frac{1}{2-n_F}} = 0 \qquad (46)$$

and by substitution of eqns. (39), (42), (44) and (45) (see Table 2.4) in eqn. (29), thus

$$\{K_A(\Delta P_{CF} + \Delta P_W)\}^{\frac{1}{2-n_A}} + \{K_E(\Delta P_{CF} + \Delta P_W)\}^{\frac{1}{2-n_E}} + (K_{WB}\,\Delta P_W)^{\frac{1}{2-n_{WB}}} + (K_{WBP}\,\Delta P_W)^{\frac{1}{2-n_{WBP}}} - 1 = 0 \qquad (47)$$

The nodal mass balance for either end region is given by eqn. (47) again and by the substitution of eqns. (39)–(43) (see Table 2.4) in eqn. (30) thus

$$\tfrac{1}{2}\{K_A(\Delta P_{CF} + \Delta P_W)\}^{\frac{1}{2-n_A}} + (K_B\,\Delta P_{CF})^{\frac{1}{2-n_B}} + (K_C\,\Delta P_{CF})^{\frac{1}{2-n_C}} + \tfrac{1}{2}\{K_E(\Delta P_{CF} + \Delta P_W)\}^{\frac{1}{2-n_E}} + (K_F\,\Delta P_{CF})^{\frac{1}{2-n_F}} - 1 = 0 \qquad (48)$$

Thus eqns. (46) and (47) for a middle region and eqns. (47) and (48) for an end region are solved by iterating on the crossflow and window zone pressure drops ΔP_{CF} and ΔP_W respectively. Coefficients K and exponents n in these equations are updated each time round the loop. This requires that the flow fractions and hence Reynolds numbers are calculated each time. The Newton–Raphson technique is used to obtain convergence and the iteration is stopped when successive values of both crossflow and window zone pressure drop are within 1% of each other.

The accuracy of the above method depends very much on the selection of values for K and n for the different flow paths. These are not readily available because they have considerable commercial value to organisations such as the Heat Transfer and Fluid Service (HTFS) and the Heat Transfer Research Institute (HTRI).

5.5. Heat Transfer Models

In the Bell[9] method the calculation of overall shell-side heat transfer coefficient is handled in a similar way to the calculation of overall pressure drop, i.e. by the calculation of a heat transfer coefficient for an ideal tube bank which is then corrected for by-passing and leakage. That is

$$h_o = h_I F_b F_l F_w F_n \qquad (49)$$

where,

- h_I is the heat transfer coefficient for crossflow through an ideal tube bank ($W/m^2\,K$),
- F_b is the correction factor for by-passing,
- F_l is the correction factor for baffle leakage,
- F_w is the correction factor for baffle window geometry, and
- F_n is the correction factor for the effect of number of tube rows on heat transfer rate.

Correction factor F_w takes account of the difference in heat transfer coefficient in a window zone compared to that in a crossflow zone.

In the Tinker type methods described in Section 5.4 the fractions of the total flow through the different parts of the tube bundle are obtained implicitly in the calculation of the pressure drop. The calculation of crossflow zone heat transfer coefficient is then based on the calculated flow through the tube bundle. This consists of tube bundle flow plus a contribution from the other flows which depends on the type of flow, e.g. tube/baffle leakage flow will be more effective for heat transfer than baffle/shell flow. A similar procedure is followed for the window and end zones. In the window zone, the flow is part in crossflow and part in longitudinal flow. The overall coefficient is finally obtained by weighting the crossflow, window and end zone coefficients according to the fraction of total tube surface area in these zones.

5.6. Heat Transfer Coefficient

The Nusselt number for a tube bank with 10 or more rows is given in an ESDU publication.[12] This is

$$\mathrm{Nu} = \mathrm{Nu}'_R \beta_5 \beta_6 \tag{50}$$

where Nu'_R is the Nusselt number at a reference condition. Factors β_5 and β_6 are correction factors and correct Nu' for the following effects:

Factor	*Description*
β_5	Allows for variation of fluid properties due to $T_W \neq T_b$
β_6	Allows for tube inclination to crossflow

Values of Nu'_R are obtained using the equation

$$\mathrm{Nu}'_R = a\,\mathrm{Re}_b^m \mathrm{Pr}_b^{0{\cdot}34} \tag{51}$$

TABLE 2.5
VALUES OF a AND m TO BE USED IN EQN. (51)

Staggered tube arrangement

Re_{bm} range	X_t	X_l	a	m
1×10^1–3×10^2	All values		1·309	0·360
3×10^2–2×10^5	between		0·273	0·635
2×10^5–2×10^6	0·6 and 4		0·124	0·700

In-line tube arrangement

Re_{bm} range	X_t	X_l	X_t/X_l	a	m
1×10^1–3×10^2				0·742	0·431
3×10^2–2×10^5	1·2–4	≥1·15	≥0·06	0·211	0·651
2×10^5–2×10^6				0·116	0·700

The above values also apply for $X_t/X = 1{\cdot}0$ and $Z_t \geq 1{\cdot}05$.

In-line arrangement (less common tube spacings)

Re_{bm} range	X_t	X_l	X_t/X_l	a	m
2×10^2–2×10^5	≥1·2		0·4–0·5	0·291	0·609
	1·05–1·1		0·35–0·55	0·177	0·643
	4–6	1·4–1·7		0·223	0·626
	2–3	1·1–1·15		0·107	0·708
	4–6	1·1–1·2		0·0624	0·730
	1·1	1·0		0·206	0·639
	1·25	1·0		0·123	0·679
	1·5	1·0		0·0834	0·707
	2·0	1·0		0·0593	0·727
	4–6	1·0		0·0435	0·744

where Re and Pr are calculated at bulk fluid conditions. Equation (51) is evaluated for in-line and staggered tube arrangements using the values of a and m given in Table 2.5.

6. CHOICE OF UNIT

The heat exchanger illustrated in Fig. 2.2 is a TEMA-AES-type where E refers to conditions on the shell side. There are a number of other types which can be used, for example F-, G-, H- and J-types as in Fig. 2.1. For a

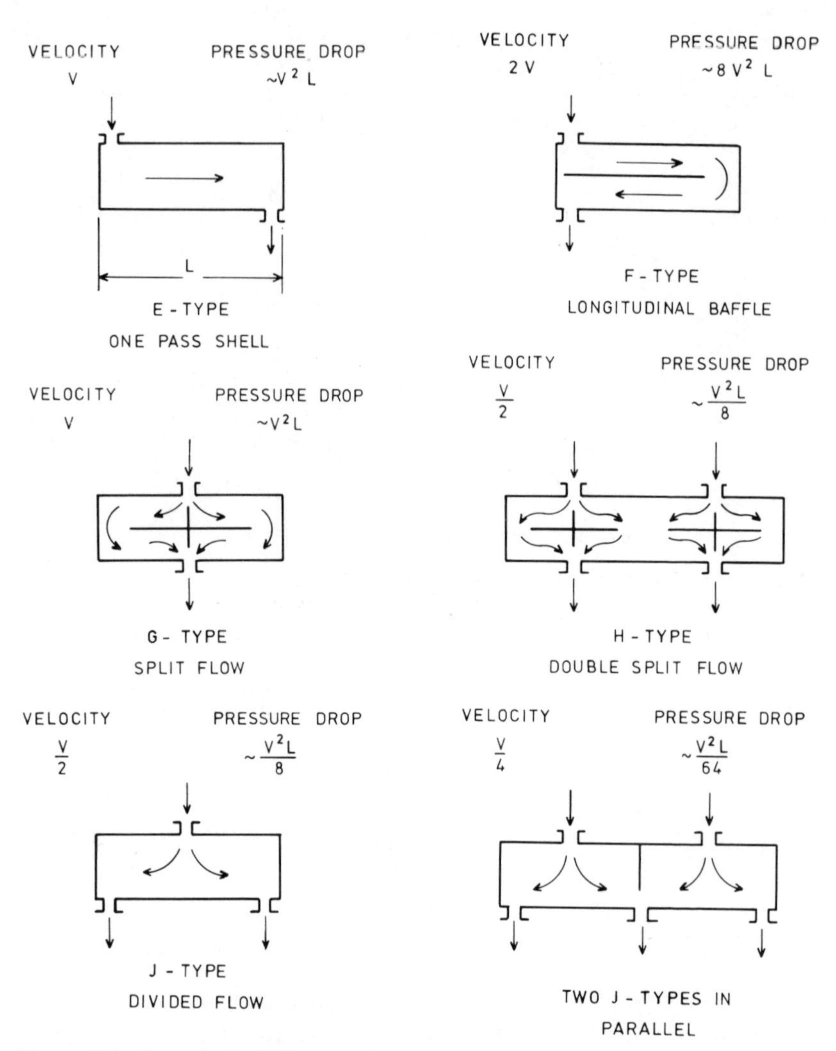

FIG. 2.10. Forms of baffled shell-and-tube heat exchanger. Type refers to conditions on shell side. (After Chisholm.)[13]

given size of shell and mass flow rate there is a considerable variation in shell-side pressure drop depending on the type of shell used. A useful illustration[13] of this is shown in Fig. 2.10 where taking an E-type shell as a reference, the pressure drop for an F-type is eight times greater and for a J-type eight times less than for the E-type.

6.1. Baffle Types

Baffles are installed on the shell side to give a higher heat transfer rate due to increased turbulence, and to support the tubes thus reducing the chance of damage due to vibration.

Plate Baffles

There are a number of different baffle types which give this turbulence due to crossflow, i.e., flow at right angles to the tubes. These are called plate baffles and are illustrated in Fig. 2.11.

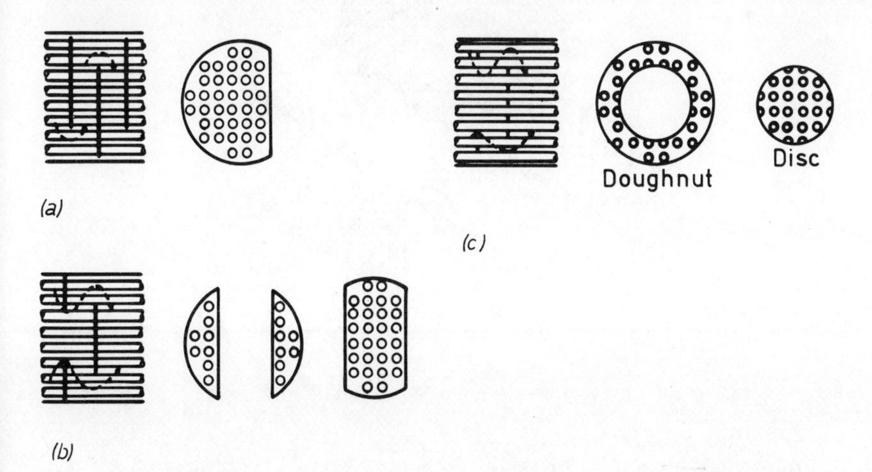

FIG. 2.11. Plate baffles giving crossflow. (a), Single segmental baffle; (b), double segmental baffle; (c), disc-and-doughnut baffle.

Another plate baffle called the orifice baffle gives longitudinal flow. Here there is sufficient clearance between the tube and baffle hole to allow longitudinal flow. The plate in this case occupies the full shell. The turbulence for high heat transfer occurs as the flow crosses the baffle.

Rod Baffles

The rod baffle design originated from work in Phillips Petroleum Co.[14] to design a tube bundle which would be free from tube vibration failure. This need resulted in a design where tubes are supported in all four directions by a rod matrix. The rod diameter is equal to the spacing between the tubes. Different arrangements of baffle were tried at the development stage and these are shown in Fig. 2.12.

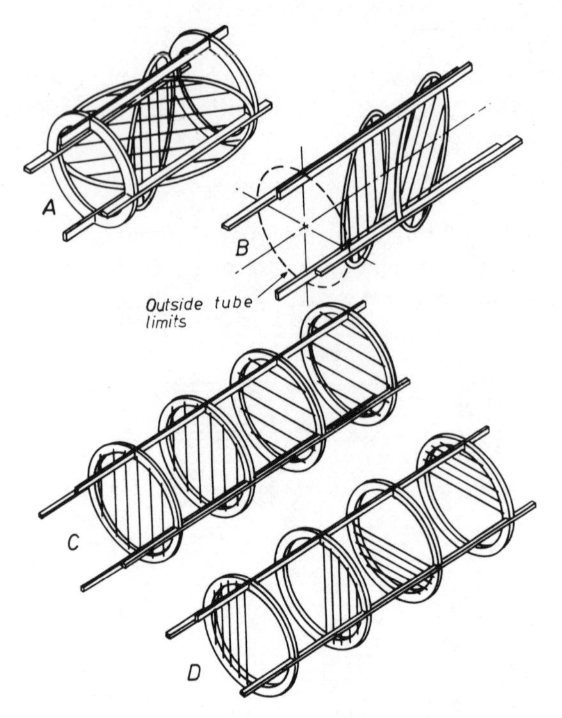

FIG. 2.12. Rod–baffle designs. (After Small and Young.)[14]

Figures 2.12(c) and 2.12(d) show the later designs where four baffles (two horizontal and two vertical) are needed to support the tubes on four sides. This is called a baffle-set and the design shown in Fig. 2.12(c) is now the preferred one; in this design, alternating tube rows are supported by rods in successive baffles.

6.2. Baffle Performance

Small and Young[14] use the ratio of overall heat transfer coefficient to shell-side pressure drop to define heat exchanger efficiency. This allows a comparison between a double-segmental baffle design and two rod baffle designs as shown in Fig. 2.13.

The rod baffle designs are referred to as a criss-cross design, Fig. 2.12(a), and vertical design, Fig. 2.12(d). The preferred 'alternate rod baffle', is not shown in Fig. 2.13 but a similar performance to that of Fig. 2.12(d) is expected. On this basis the rod baffle efficiency is approximately 83% higher than the plate baffle efficiency.

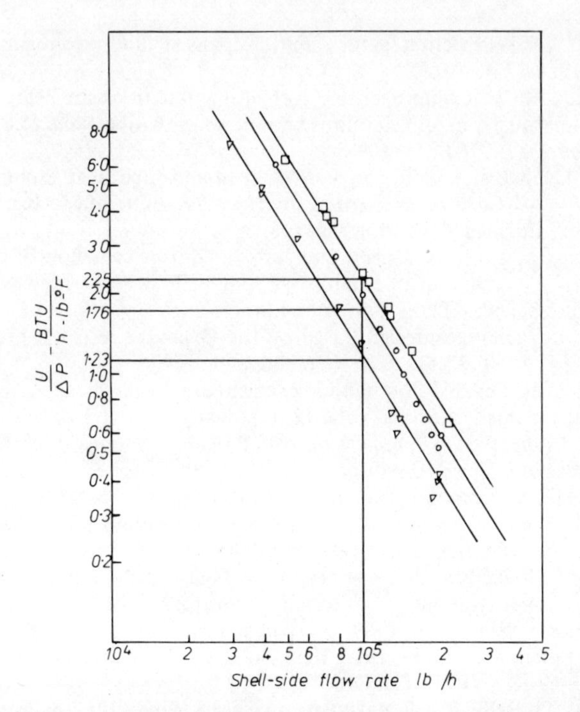

FIG. 2.13. Ratio of heat transfer to pressure loss. ▽ Double-segmental plate baffle. Baffle spacing = 4·9 in; □ vertical rod baffle. Baffle spacing = 2·4 in; ○ criss-cross rod baffle. Baffle spacing = 9·8 in. (After Small and Young.)[14]

REFERENCES

1. *Standards of the Tubular Exchanger Manufacturers Association* (*TEMA*), 5th ed., 1968, Tubular Exchanger Manufacturers Association, New York.
2. EMERSON, W. H. Effective tube-side temperatures in multi-pass heat exchangers with non-uniform heat transfer coefficients and specific heats, in: *Advances in Thermal and Mechanical Design of Shell-and-Tube Heat Exchangers*, 1975, Report of a meeting at NEL, 28 November 1973, NEL Report No. 590. National Engineering Laboratory, East Kilbride, Glasgow.
3. BUTTERWORTH, D. A calculation method for shell-and-tube heat exchangers in which the overall coefficient varies along the length, in: *Advances in Thermal and Mechanical Design of Shell-and-Tube Heat Exchangers*, 1975, Report of a meeting at NEL, 28 November 1973, NEL Report No. 590. National Engineering Laboratory, East Kilbride, Glasgow.
4. JAW, W. Temperature relations in shell-and-tube heat exchangers having one-pass split flow shells, *J. Heat Trans., Trans. ASME*, 1964, **C86,** 408.

5. CHURCHILL, S. W. Friction factor equation spans all fluid-flow regimes, *Chem. Eng.*, 1977, **84**(24), 91–92.
6. CHURCHILL, S. W. Comprehensive correlating equations for heat, mass and momentum transfer in fully developed flow in smooth tubes, *Ind. Eng. Chem.—Fundamentals*, 1977, **16**(1), 109–15.
7. TINKER, T. Shell-side characteristics of shell-and-tube heat exchangers, in: *Proceedings of General Discussion on Heat Transfer*, 1951, Institution of Mechanical Engineers, London, 89–116.
8. ENGINEERING SCIENCE DATA UNIT. Pressure loss during crossflow of fluids with heat transfer over plain tube banks without baffles, 1974, Item No. 74040, Engineering Sciences Data Unit, London.
9. BELL, K. J. Exchanger design based on the Delaware research programme, *Petrol. Eng.*, 1960, **32**(11), C26–C36 and C40a–C40c.
10. GRANT, I. D. R. Divided-flow method of calculating shell-side pressure drop for segmentally baffled shell-and-tube heat exchangers. *CAD 74, Proc. of Int. Conf. on Computers in Engineering and Building Design*, 25–27 September 1974, Imperial College, London.
11. GRANT, I. D. R. Flow and pressure drop with single-phase and two-phase flow on the shell side of segmentally baffled shell-and-tube heat exchangers, in: *Advances in Thermal and Mechanical Design of Shell-and-Tube Heat Exchangers*, 1975, Report of a meeting at NEL, 28 November 1973, NEL Report No. 590, National Engineering Laboratory, East Kilbride, Glasgow.
12. ENGINEERING SCIENCE DATA UNIT. Convective heat transfer during crossflow of fluids over plain tube banks, 1973, Item No. 73031, Engineering Sciences Data Unit, 251–59 Regent St., London.
13. CHISHOLM, D. Baffled shell-and-tube condensers, *Notes for Condensation and Condensers Course*, 11–15 September 1972. Birniehill Institute, National Engineering Laboratory, East Kilbride, Glasgow.
14. SMALL, W. M. and YOUNG, R. K. Exchanger design cuts tube vibration failures, *Oil Gas J.*, 1977, **75**(37), 77–80.

Chapter 3

REBOILERS

G. T. Polley

National Engineering Laboratory, East Kilbride, Glasgow, UK

SUMMARY

The various ways in which shell-and-tube heat exchangers can be used to provide boil-up at the base of a distillation column are considered. The various reboiler types (kettle, vertical thermosiphon, horizontal thermosiphon and forced recirculation) are described and the factors that influence their selection discussed.

The thermal and hydraulic processes that occur within shell-and-tube heat exchangers on such duties are described. Equations and procedures for the estimation of heat transfer coefficient, pressure drop and two phase flow pattern are presented and discussed.

Strategies for the design of the various types of reboiler are proposed.

NOTATION

A	Area (m^2)
B	Chisholm's B coefficient
C	Chisholm's C coefficient
c	Heat capacity (J/kg K)
D_B	Tube bundle diameter (eqn. (31)) (ft)
d	Tube diameter (m)
F	Chen's F factor
f	Friction factor
g	Gravitational acceleration (m/s^2)

h_v	Enthalpy of vaporisation (J/kg)
l	Length (m)
$\dot{m}$	Mass flux (kg/m^3)
P	Pressure (N/m^2)
P_t	Tube pitch (eqn. (31)) (ft)
Q	Heat flow (W)
S	Chen's S factor
T	Temperature (K)
U^s	Superficial velocity (m/s)
X_{tt}	Lockhart–Martinelli parameter
x	Mass quality

Greek symbols

α	Heat transfer coefficient (W/m^2 K)
ε_v	Void fraction
ρ	Density (kg/m^3)
η	Viscosity (Ns/m^2)
ϕ	Tube bundle geometry factor (eqn. (31))
λ	Thermal conductivity (W/m K)
ψ	Physical property factor (eqn. (32)) (Btu/ft^3 h)
σ	Surface tension (N/m)
θ	Inclination (radians)
Γ	Baroczy index (eqn. (21))

Dimensionless groups

Pr	Prandtl number ($c\mu/\lambda$)
Re	Reynolds number ($\dot{m}d/\mu$)
St	Stanton number ($\alpha/\dot{m}c$)

Subscripts

A	Accelerational component
B	Boiling component
B*i*	Boiling incipience
c	Critical
F	Frictional
FC	Forced convection
G	Gas
G	Gas Fraction flowing alone
GO	Total mass flowing as gas
hom	Homogeneous

L	Liquid
L	Liquid fraction flowing alone
LO	Total mass flowing as liquid
ONB	Onset of nucleate boiling
sat	Saturation
TP	Two phase

1. INTRODUCTION

Reboiler is the name given to the heat exchanger used to provide boil-up at the base of a distillation column. This chapter considers the various ways in which shell-and-tube heat exchangers are used as reboilers. The discussion of the thermal and hydraulic processes occurring within such equipment will be restricted to the cold (i.e. vaporisation) side.

All of the reboiler types considered employ a flow boiling process. They differ in the manner in which this flow is promoted (forced or natural circulation), in the side of the exchanger used for vapour generation (shell or tube), and in the detailed construction of the exchanger (shell type).

Shell types will be referred to by their TEMA[1] designation. The principal shell types used in this application are E, G, H and K. See Fig. 2.1.

2. BOILER CONFIGURATIONS

2.1. Kettle Reboilers

The kettle reboiler consists of a bundle of heated tubes situated in an oversized shell (usually a TEMA K-type). Tube bundle to shell diameter ratios typically fall in the range 1·3–2·0, and are set by disentrainment considerations. The tube bundles are usually of circular cross-section. However, in some circumstances semi-circular bundles occupying the whole of the lower half of the shell may be used.

A typical kettle reboiler installation at the base of a distillation column is shown in Fig. 3.1(a). The feed to the reboiler is at a rate that is greater than the vaporisation rate in order to prevent the build up of low volatility materials in the shell. Excess liquid flows over a weir situated near the end of the tube bundle. (Weir heights are typically between 5 and 15 cm greater than the bundle height.) On occasions in which the bottom product rate from the column is significantly lower than the vaporisation rate, excess liquid flow over the weir should be promoted and a return to the column

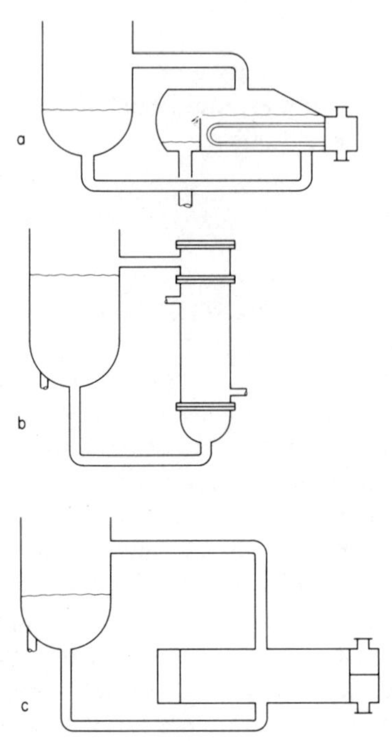

FIG. 3.1. (a) Kettle reboiler installation; (b) vertical thermosiphon reboiler installation; (c) horizontal thermosiphon reboiler installation.

provided. The throughput that should be used is dependent upon bottom composition and other process engineering considerations.

Although the kettle reboiler is often considered to be a 'pool boiling' device, it actually relies upon a flow boiling mechanism. A natural circulation loop is set up within the shell as flow of liquid around and through the bundle is induced by the difference between the density of the liquid surrounding the bundle and the two-phase mixture generated within the bundle. There is a marked variation in the quality (vapour fraction) of the two-phase mixture over the depth of the bundle. This gives rise to a marked variation in heat transfer over the depth of the bundle (Fig. 3.2).[2]

The maximum heat flux that may be accommodated by a kettle reboiler bundle before partial dryout (with associated fall-off in heat transfer coefficient) occurs, is set by the liquid recirculation rate.[3] Consequently, fairly open bundles are preferred and tube pitch to diameter ratios are usually in the range 1·5–2. This maximum heat flux will be referred to as 'limiting heat flux'.

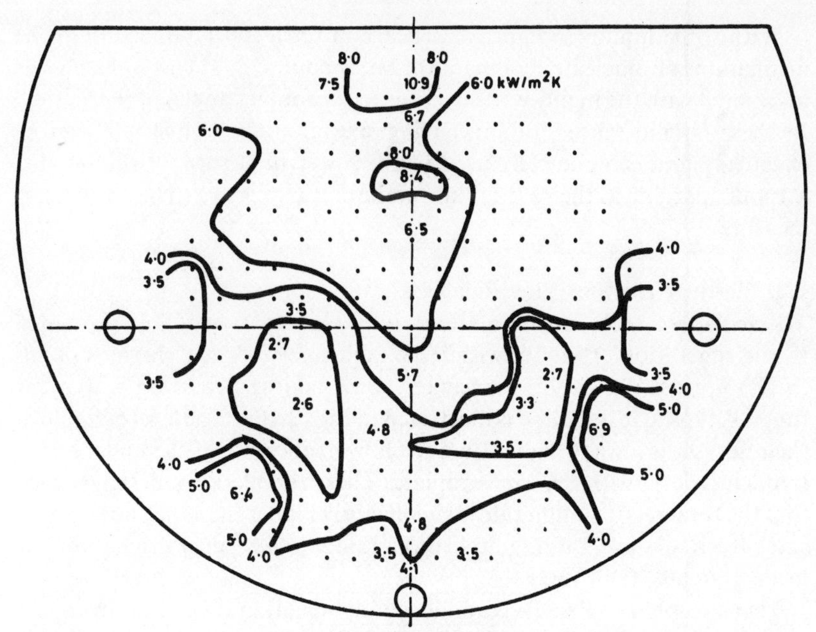

FIG. 3.2. Distribution of heat transfer coefficient in a kettle reboiler tube bundle.

Long thin bundles have higher limiting heat fluxes than short fat ones. However, short fat bundles have higher heat transfer coefficients whilst within operational range.

Vapour and liquid are almost completely separated within the kettle shell. Any liquid that is present in the vapour stream is usually the result of entrainment. Disentrainment devices (e.g. mesh pads) are frequently incorporated at the vapour outlet port of this type of reboiler.

The kettle reboiler is not very suitable for use with liquids that have a tendency to foam. Reboilers that generate thin films, as opposed to bubbly mixtures, are more suitable for such duties.

Residence times in kettle reboilers are fairly high. Consequently, kettle reboilers are not suitable for use with heat-sensitive materials.

Flow velocities in these reboilers tend to be lower than those achieved with forced circulation, or in vertical thermosiphon reboilers of good design. Consequently, these reboilers may be expected to have a greater tendency to foul. However, kettle reboilers are relatively easy to maintain.

The large shell size for a given bundle size often means that kettle reboilers are not suitable (on economic grounds) for high-pressure applications.

With some liquids (notably fluorocarbon refrigerants) difficulty in the instigation of nucleate boiling can be encountered. This difficulty is associated with the highly wetting nature (low contact angle) of the liquid.[4] In these circumstances, tubing having special surface finishes[5,6] can be effectively and economically used in the lower tube rows of the bundle. Vapour sparging at the base of the bundle can also prove to be an effective measure.

2.2. Vertical Thermosiphon Reboilers

Thermosiphon reboilers rely on the natural recirculation of fluid promoted by the generation of vapour within the exchanger. Vertical thermosiphons (VTR) are usually E-type exchangers with boiling occurring within the tubes. A typical installation is illustrated in Fig. 3.1(b). The level of liquid in the reservoir is always higher than the lower tubeplate of the boiler and is frequently level with the top tubeplate. These reboilers are designed such that the total recirculation rate is significantly higher than the vaporisation rate. Recirculation ratios (circulation rate/vapour generation rate) are usually greater than three.

Thermosiphon reboilers cannot be designed in isolation from the associated pipework. The outlet pipework must be as free from restriction as possible since such restrictions not only reduce recirculation but can also induce flow instability. It is common practice to make the flow area of the outlet piping equal to that of the boiler tubes. Restrictions in the boiler feed piping improve flow stability and are often deliberately incorporated. The problem of flow stability is outside of the scope of this work. A useful discussion of this problem has been made by Bailey.[7]

A vertical configuration means that the VTR is economical in terms of ground space and support structure. However, it does mean that the distillation column must be elevated above the ground and it does make exchanger maintenance costly.

High recirculation rates make the VTR less prone to fouling than some other types. It also gives the reboiler relatively low residence time.

2.3. Horizontal Thermosiphon Reboilers

In horizontal thermosiphon reboilers the boiling occurs on the shell side of the exchanger. A typical installation is shown in Fig. 3.1(c). The shells are usually either G-type or H-type. The outlet nozzle may sometimes be replaced by a dome extending some distance along the top of the shell. The longitudinal baffle is usually solid but perforated baffles are occasionally used. In some designs, baffling is only used in the lower (low quality) half of the shell. As with the boiler types discussed above, the circulation through the boiler is significantly higher than the vaporisation rate.

The choice of which thermosiphon type (VTR or HTR) to be used for a given duty is dependent upon several factors. The larger the surface area requirement, the more the horizontal unit is favoured (because of height and weight limitations). If the liquid being vaporised has a tendency to foul, a vertical system may be favoured (tubes being easier to clean than shells). Similarly, if the heating medium has a tendency to foul, the horizontal unit will be favoured. It is usually easier to maintain a horizontal unit than a vertical unit. Horizontal units have a lower boiling point elevation than vertical units and consequently are more suited to vacuum operation.

2.4. Forced Circulation Reboilers

Forced recirculation reboilers are similar in most respects to thermosiphon reboilers, the difference being that the liquid is pumped through the exchanger (which makes them easier to design than thermosiphon systems). In many cases the pump used to extract the bottom product from a distillation column is used to provide the flow through the boiler. Consequently, the choice between natural and forced circulation sometimes only involves the sizing of mechanical equipment and not the provision of extra equipment.

Forced circulation systems are preferred in cases in which fouling is likely to occur (the fall-off in flow with increased flow resistance is generally less marked), where viscous or solid-bearing fluids are handled, and in cases in which the vaporisation rate is low.

3. HEAT TRANSFER AND PRESSURE DROP IN REBOILERS

3.1. Flow Boiling in Vertical Tubes

When an initially subcooled liquid flows through a heated vertical tube, the walls of which are maintained at a temperature higher than the local saturation temperature of the liquid, the way in which heat is transferred varies along the length of the tube in a manner that is related to the changing structure of the two-phase flow pattern within the tube. A possible variation of two-phase flow pattern along the length of a tube is illustrated in Fig. 3.3.

At the entrance to the tube the subcooling of the liquid is often such that nucleation (formation of vapour bubbles at the solid surface) does not occur and heat is transferred by single-phase convection. As the temperature of the fluid increases there comes a point at which nucleation

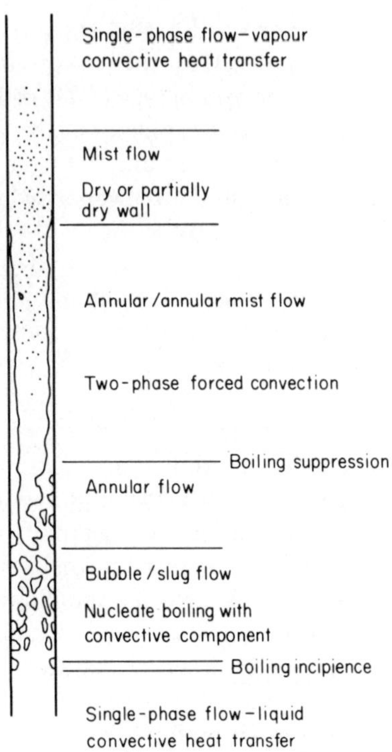

FIG. 3.3. Variation of a two-phase flow pattern along a vertical tube.

does occur (boiling incipience) and heat is subsequently transferred by subcooled nucleate boiling. As more heat is added the liquid becomes saturated, and saturated flow boiling develops. With increasing vapour generation the structure of the two-phase mixture changes from bubbly flow to slug flow (large slugs of liquid and vapour interspersed, each of cross-section near to that of the tube), then from slug flow to annular flow. With annular flow the liquid is dispersed as a thin film at the tube wall (with some entrained as droplets in the gas core). Forced convective heat transfer in annular flow is very efficient. This results in a reduction in the tube wall temperature and frequently a point is reached at which nucleation is suppressed and heat is transferred solely by convection. As the quality of the stream is increased further, the thickness of the liquid film is reduced by evaporation and by entrainment. Eventually dry patches are formed on the tube wall. The formation of these dry patches results in a radical reduction in heat transfer coefficient. This area of the tube is said to be in the post-dryout region.

The annular flow regime sets in at low qualities (typically 1–2%), and is the most important of the flow regimes. Fortunately it is a regime in which we can predict heat transfer with confidence.

Designers should avoid designing reboilers which operate close to or beyond the dryout point. Consequently, only two regimes need be considered for a reasonable design approximation. These are the subcooled and the saturated annular flow regimes. These regimes may require subdivision into nucleate and convective zones.

Single-phase forced convection in the turbulent flow regime is well predicted by the ESDU[8] correlation:

$$\mathrm{St} = \exp\{-3{\cdot}796 - 0{\cdot}205\ln \mathrm{Re} - 0{\cdot}505\ln \mathrm{Pr} - 0{\cdot}0225(\ln \mathrm{Pr})^2\} \quad (1)$$

The boundary between the single-phase convective zone and the subcooled nucleate boiling subzone is dependent upon heat flux, bulk temperature and the single-phase convective heat transfer coefficient, and can be calculated from the equation for boiling incipience developed by Davis and Anderson:[9]

$$\left(\frac{Q}{A}\right)_{\mathrm{ONB}} = \frac{\lambda\,\Delta h_{\mathrm{v}}\rho_{\mathrm{v}}}{8\sigma T_{\mathrm{sat}}}(\Delta T_{\mathrm{sat}})^2_{\mathrm{ONB}} \quad (2)$$

This equation can also be used for the determination of the boundary between the nucleate boiling and two-phase convective zones with saturated flow boiling.

Methods for the correlation and prediction of subcooled flow boiling fall into three categories: superposition techniques, dimensionless group correlations, and interpolation techniques.

The Chen[10] correlation (which is described below for saturated flow boiling) can be assumed to be a superposition technique. This correlation was derived for saturated flow boiling. However, an attempt has been made to extend it to subcooled boiling duties without success.[11]

The dimensionless group technique has been successfully employed by Moles and Shaw[11] for the correlation of data from 10 sources relating to seven liquids and six surfaces. A total of 664 experimental measurements were correlated to an accuracy of $\pm 40\%$ by the equation:

$$\frac{\alpha}{\alpha_{\mathrm{FC}}} = 78{\cdot}5\left(\frac{Q/A}{\Delta h_{\mathrm{v}}\rho_{\mathrm{v}}U^{\mathrm{s}}}\right)^{0\cdot67}\left(\frac{\Delta h_{\mathrm{v}}}{C\,\Delta T_{\mathrm{sub}}}\right)^{0\cdot5}\left(\frac{\rho_{\mathrm{v}}}{\rho_{\mathrm{L}}}\right)^{0\cdot7}\mathrm{Pr}_{\mathrm{L}}^{0\cdot46} \quad (3)$$

In this equation the Prandtl number was evaluated at the mean film temperature whilst all other groups were evaluated at the saturation temperature.

The best available technique is probably that proposed by Bergles and Rohsenow,[12] which utilises the interpolation technique. Their equation is:

$$\frac{Q}{A} = \left(\frac{Q}{A}\right)_{FC}\left[1 + \left[\frac{(Q/A)_B}{(Q/A)_{FC}}\left[1 - \frac{(Q/A)_{Bi}}{(Q/A)_B}\right]\right]^2\right]^{1/2} \tag{4}$$

where $(Q/A)_{Bi}$ is the boiling heat flux associated with the temperature driving force required for boiling incipience. The predictions of this equation closely follow measurements of subcooled boiling heat transfer

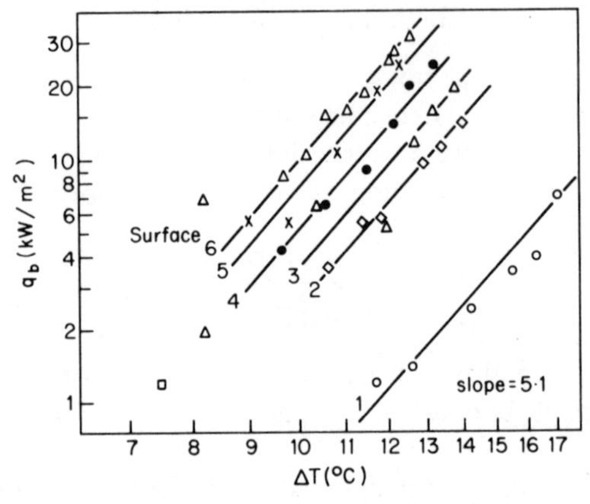

FIG. 3.4. The effect of surface on nucleate boiling heat transfer.

with water[12] and with R-113.[13] The problem in the application of the method is the difficulty of predicting nucleate boiling heat transfer for a surface in the absence of experimental data. This is a problem associated with all attempts to obtain a general equation for processes involving nucleate boiling. Boiling heat transfer is a very strong function of the surface characteristics of the hot wall. An example of this dependency is shown in Fig. 3.4, in which data relating to the boiling of water on copper plates of different roughnesses are presented.

Best predictions from eqn. (4) are obtained if the pool boiling component is based upon experimental data for the actual surface under consideration. Of course this is rarely possible. The next resort is to a correlation based upon experimental data for the given fluid/surface type under consideration (e.g. Voloshko's[14] correlations for refrigerants boiling upon a stainless steel surface). As a last resort one of the general correlations may be used. The general correlation recommended by Palen *et al.*[15] is that proposed by Mostinski:[16]

$$\alpha_{PB} = 0{\cdot}006\,58\mathrm{P}_c^{0{\cdot}69}\left(\frac{Q}{A}\right)^{0{\cdot}7}\left\{1{\cdot}8\left(\frac{\mathrm{P}}{\mathrm{P}_c}\right)^{0{\cdot}17} + 4\left(\frac{\mathrm{P}}{\mathrm{P}_c}\right)^{1{\cdot}2} + 10\left(\frac{\mathrm{P}}{\mathrm{P}_c}\right)\right\}^{10} \quad (5)$$

Stephan and Abdelsalam[37] have recently published correlations for pool boiling heat transfer that have been obtained by applying regression analysis to a large quantity of data extracted from over 50 reports. One equation that covers all substances was presented. The absolute mean error found with this equation was 22·3%. Better accuracy was obtained by correlating data for specific chemical types. Equations are reported for water (mean absolute error 11·3%), hydrocarbons (12·2%), cryogenic fluids (14·3%) and refrigerants (10·6%).

Several empirical correlations have been suggested for the estimation of heat transfer during saturated flow boiling in vertical tubes (see review by Hsu and Graham).[17] The most widely used and accepted correlation to date is that of Chen.[10] He suggests that the heat transfer coefficient is made up of two additive components:

$$\alpha = F\,.\,\alpha_{FC} + S\,.\,\alpha_{NB} \quad (6)$$

where, α_{NB} is the nucleate boiling coefficient with zero flow; S is a 'suppression' factor; α_{FC} is the single-phase forced convection heat transfer coefficient if the liquid content of the mixture were flowing alone; and F is a two-phase convection enhancement factor.

The product $F\alpha_{FC}$, is actually the two-phase forced convection heat transfer coefficient. The factor F is well correlated by the equation:

$$F = \left(\frac{\Delta\mathrm{P}_{TP}}{\Delta\mathrm{P}_L}\right)^{0{\cdot}89} \quad (7)$$

which Chen derived from a heat transfer/momentum analogy. Two-phase pressure drop is described below. Substitution of eqn. (17) into eqn. (7) yields:

$$F = \left[1 + \frac{\left(\frac{\rho_L}{\rho_v}\right)^{1/2} + \left(\frac{\rho_v}{\rho_L}\right)^{1/2}}{X_{tt}} + \frac{1}{X_{tt}^2}\right]^{0{\cdot}445} \quad (8)$$

A comparison of the predictions of this equation with the data of Guerrieri and Talty[18] is made in Fig. 3.5. It is seen that the equation accounts for the effect of differing physical properties on heat transfer. Chen presented values of F as a graphical relationship against $1/X_{tt}$ (Fig. 3.6). A

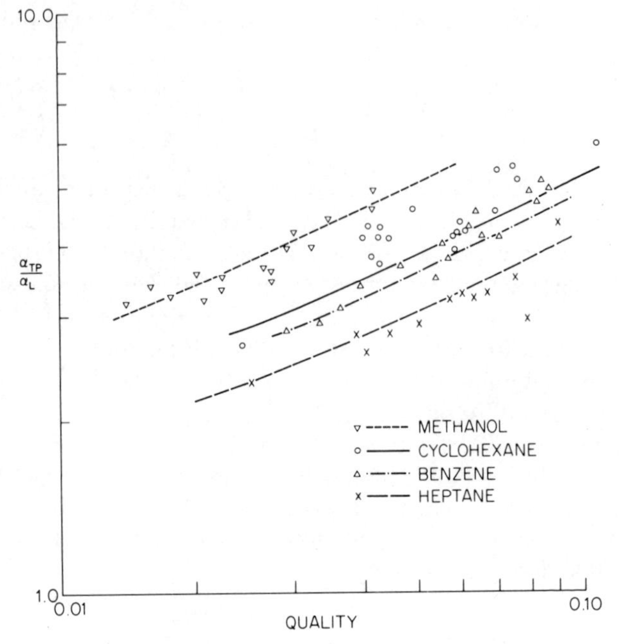

FIG. 3.5. Two-phase forced convective boiling in tubes.

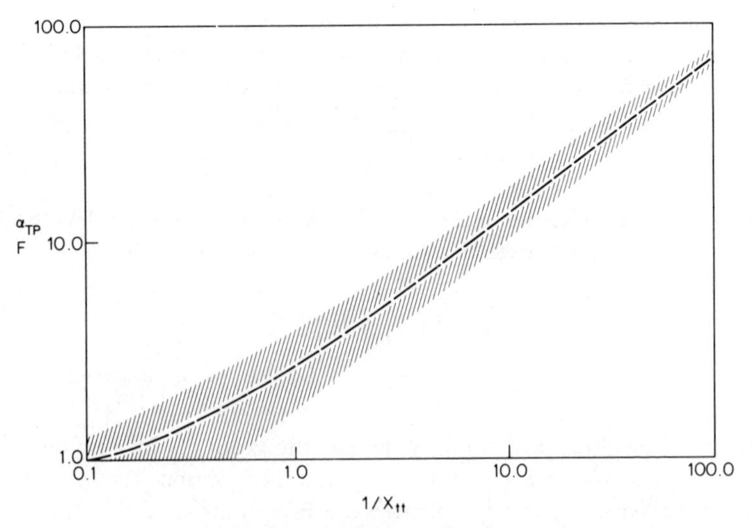

FIG. 3.6. Chen's F factor.

comparison of Figs. 3.5 and 3.6 suggests that the large scatter of experimental data about the curve at low values of $1/X_{tt}$ may be a function of the physical properties of the fluid.

Chen also reports the suppression factor S in a graphical form (Fig. 3.7).

The weakness of the Chen method is again the evaluation of the nucleate boiling term. Chen assumed that the nucleate boiling term was given by the Forster–Zuber equation:

$$\alpha_{PB} = 0{\cdot}001\,22\left[\frac{\lambda_L^{0\cdot79} c_L^{0\cdot45} \rho_L^{0\cdot49}}{\sigma^{0\cdot5}\,\Delta h_v^{0\cdot24} \mu_L^{0\cdot29} \rho_v^{0\cdot24}}\right]\Delta T_c^{0\cdot24}\,\Delta P_v^{0\cdot75} \tag{9}$$

Since, the suppression factors are all based upon this equation it is advisable to assume that this equation holds for all surfaces when applying this equation.

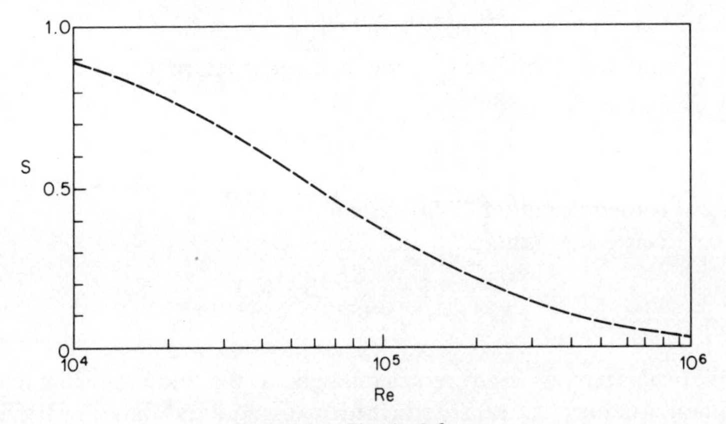

FIG. 3.7. Chen's S factor.

For cases in which nucleation is suppressed and heat is transferred by convection alone, eqn. (8) reduces to:

$$\alpha = F\,.\,\alpha_{FC} \tag{10}$$

The designer must consider both pressure drop and heat transfer. In the case of pressure drop, the basic equations of two-phase flow have been described and discussed by Collier.[19] The pressure drop in a two-phase system is made up of three components. One component is associated with friction, one with gravity, and the third with the acceleration of the fluids. The overall pressure drop is the summation of these terms. Two basic models exist. These are the 'homogeneous' model, in which the two-phase mixture is assumed to behave as an equivalent single-phase fluid, and a 'separated flow' model in which each fluid can behave in its own manner.

With the 'homogeneous' model the frictional pressure loss is given by:

$$-\left(\frac{\partial P}{\partial l}\right)_F = 2\frac{f_{TP}}{d}\cdot\frac{\dot{m}^2}{\rho_{TP}} \tag{11}$$

The friction factor f_{TP}, is obtained from relationships for single-phase fluids but using a Reynolds number based upon a two-phase viscosity. Various definitions have been proposed for this two-phase viscosity. One of the most successful definitions is that of Dukler *et al.*[20]

$$\mu_{TP} = (1 - x_G)\mu_L + x\mu_G \tag{12}$$

The 'homogeneous' density used in the above equations is defined as:

$$1/\rho_{TP} = 1 - x_G/\rho_L + x_G/\rho_G \tag{13}$$

The gravitational component of the 'homogeneous' model is:

$$-\left(\frac{\partial P}{\partial l}\right)_g = \rho_{TP}\,.\,g\sin\theta \tag{14}$$

where θ is the inclination of the system.

The acceleration term is:

$$-\left(\frac{\partial P}{\partial l}\right)_a = \dot{m}^2\frac{\partial(1/\rho_{TP})}{\partial l} \tag{15}$$

Of particular interest to the reboiler designer is the case in which liquid is evaporated to form a stream of specified quality. Collier[19] has shown that in the case in which

(i) heat flux is uniform,
(ii) compressibility of the gas can be ignored, and
(iii) friction factor remains constant over the tube length,

the above set of equations can be integrated to yield:

$$\Delta P_{TP} = \frac{2f_{TP}l\dot{m}^2}{\rho_L d}\left\{1 + \frac{x_G}{2}\left(\frac{\rho_L}{\rho_G} - 1\right)\right\} + \dot{m}^2\left\{\frac{1}{\rho_G} - \frac{1}{\rho_L}\right\}$$
$$+ \frac{g\sin\theta l}{x_G\left(\dfrac{1}{\rho_G} - \dfrac{1}{\rho_L}\right)}\log\left\{1 + x_G\left(\frac{\rho_L}{\rho_G} - 1\right)\right\} \tag{16}$$

With the 'separated flow' model empirical, or semi-empirical, equations based upon experimental data are used for the estimation of the frictional loss. A useful equation has been proposed by Chisholm:[21]

$$\frac{\Delta P_{TP}}{\Delta P_L} = 1 + \frac{C}{X_{tt}} + \frac{1}{X_{tt}^2} \tag{17}$$

where, X_{tt} is called the Lockhart–Martinelli parameter, and is defined by:[22]

$$X_{tt} = \left(\frac{\Delta P_L}{\Delta P_G}\right)^{1/2} \tag{18}$$

where, ΔP_L is the frictional loss associated with the single-phase flow of the liquid component of the mixture flowing alone. (For the gas ΔP_G is defined in a similar manner.)

The constant C is dependent upon the physical properties of the two phases and upon the mass flux (in the application of this equation in eqn. (8) the mass flux dependency was ignored):

$$C = \left(\frac{\rho_L}{\rho_G}\right)^{1/2} + \left(\frac{\rho_G}{\rho_L}\right)^{1/2} \qquad \dot{m} \geq 2000$$

$$C = \frac{2000}{G}\left\{\left(\frac{\rho_L}{\rho_G}\right)^{1/2} + \left(\frac{\rho_G}{\rho_L}\right)^{1/2}\right\} \dot{m} < 2000 \tag{19}$$

In a later paper Chisholm[23] demonstrated how this equation could be further developed to give an equation that could be integrated with respect to quality analytically.

$$\frac{\Delta P_{TP}}{\Delta P_{LO}} = 1 + (\Gamma^2 - 1)\,[B \,.\, x_G^{(2-n)/2}(1 - x_G)^{(2-n)/2} + x_G^{2-n}] \tag{20}$$

ΔP_{LO} is the single-phase pressure drop that would be encountered if the total flow of the mixture was flowing as liquid; n is the exponent of the Reynolds number in the relationship between friction factor and Reynolds number (and is a negative number); and Γ is defined by:

$$\Gamma = \left(\frac{\Delta P_{GO}}{\Delta P_{LO}}\right)^{1/2} \tag{21}$$

Recommendations for the value of the constant B are given in Table 3.1.

With the 'separated flow' model the gravitational pressure drop is based upon the 'void fraction' (ratio of gas flow area to total flow area):

$$-\left(\frac{\partial P}{\partial l}\right)_g = g \sin\theta[\varepsilon_v \rho_G + (1 + \varepsilon_v)\rho_L] \tag{22}$$

TABLE 3.1
VALUES OF B FOR SMOOTH TUBES

Γ	$\dot{m}$	B
—	≤ 500	4·8
$\leq 9{\cdot}5$	$500 < \dot{m} < 1\,900$	$2\,400/G$
—	$\geq 1\,900$	$55/G^{0\cdot5}$
$9{\cdot}5 < \Gamma < 28$	≤ 600	$520/(\Gamma G^{0\cdot5})$
—	> 600	$21/\Gamma$
≥ 28	—	$\dfrac{15\,000}{\Gamma^2 G^{0\cdot5}}$

Many equations have been proposed for the estimation of void fraction. None of these equations gives best prediction over the full range of possible conditions. An equation which gives reasonably good predictions for most conditions is:[24]

$$\varepsilon_G = \frac{1}{1 + \left(1 - x_G + \dfrac{x_G \rho_L}{\rho_G}\right)^{1/2} \dfrac{(1 - x_G)\rho_G}{x_G \rho_L}} \tag{23}$$

Acceleration losses are given by:

$$-\left(\frac{\partial P}{\partial l}\right)_a = \dot{m}^2 \frac{\partial}{\partial l}\left[\frac{x_G^2}{\varepsilon_v \rho_G} + \frac{(1 - x_G)^2}{\rho_L(1 - \varepsilon_v)}\right] \tag{24}$$

3.2. Flow Boiling in Tube Bundles

Two-phase flow through tube bundles has been the subject of a long-term wide-ranging study at the National Engineering Laboratory, UK. The results of flow-pattern and pressure-drop studies have recently been discussed by Grant and Chisholm.[25]

The various flow patterns that have been observed with vertical and with horizontal flow are illustrated in Fig. 3.8. Grant and Chisholm describe these flow patterns as follows:

Spray flow. Liquid is carried along by the gas stream as entrained droplets (occurs at high mass flow qualities).

Bubbly flow. Gas is distributed as discrete bubbles in the liquid (occurs at low mass flow qualities).

Intermittent flow. Intermittent slugs of liquid are propelled through the bundle by the gas.

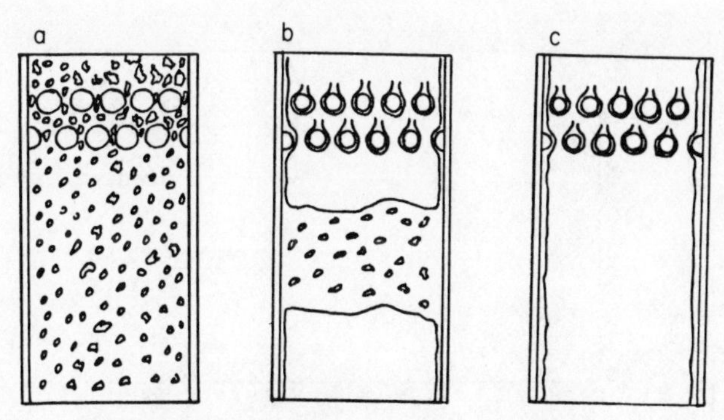

FIG. 3.8. (a). Two-phase flow patterns—vertical upward flow in tube bundles. (a) Bubbly flow; (b) slug flow; (c) spray flow.

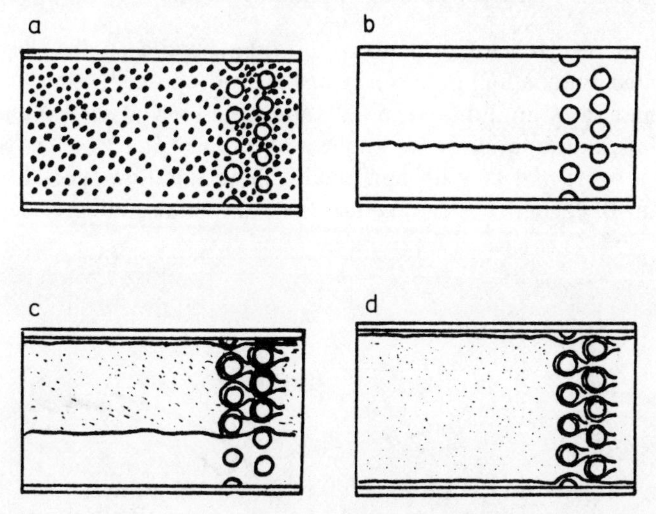

FIG. 3.8. (b). Two-phase flow patterns—horizontal flow in tube bundles. (a) Bubbly flow; (b) stratified flow; (c) stratified-spray flow; (d) spray flow.

Stratified/spray flow. Liquid and gas are tending to separate, with the liquid mainly flowing through the bottom rows of the bundle. The gas phase is partially transported as discrete bubbles in the liquid. Similarly part of the liquid is transported as discrete droplets in the gas phase.

Stratified flow. Liquid and gas are completely separated.

Flow maps for vertical flow and for horizontal flow are presented in Figs. 3.9 and 3.10.

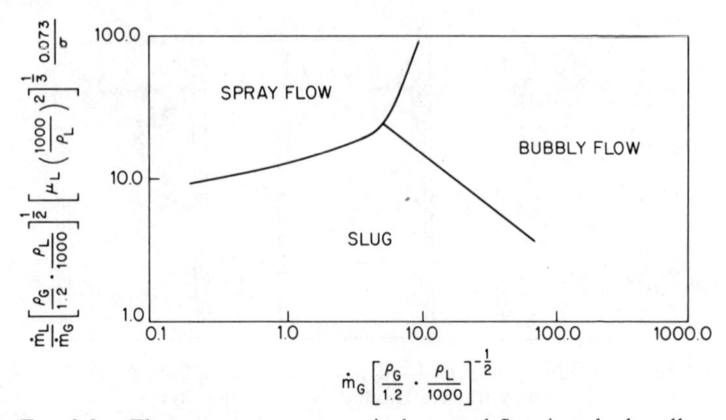

FIG. 3.9. Flow pattern map—vertical upward flow in tube bundles.

Two-phase pressure drop data for tube bundle studies have been correlated by equations of the form previously proposed by Chisholm[23] for two-phase flow in tubes (eqn. 20 above). For vertical up-and-down crossflow, data from air–water studies are well correlated when the constant B has a value of 1·0. With horizontal side-to-side flow the value of the constant B, again derived using data from air–water studies, is dependent

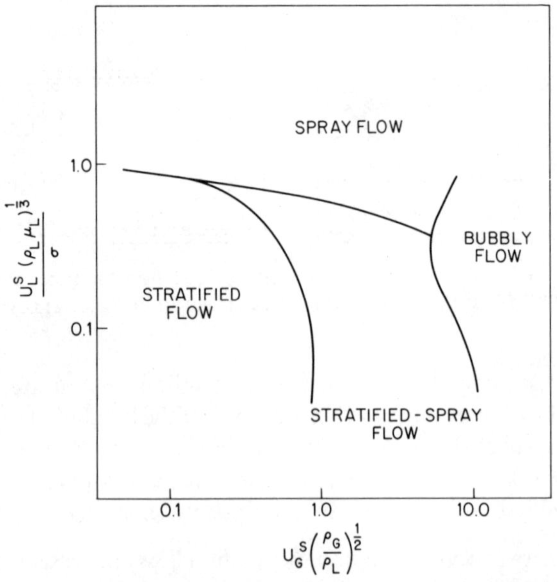

FIG. 3.10. Flow pattern map—horizontal flow in tube bundles.

upon flow regime. With stratified and stratified/spray flow the constant B has a value of 0·25. With bubbly and spray flow the constant has a value of 0·75. No dependence of B upon mass flux has yet been reported.

Pressure drop through the window zones of heat exchanger shells is usually dominated by expansion, contraction and change of direction losses and not by the frictional drag of the tubes. Under these circumstances the friction factor becomes independent of Reynolds number and eqn. (20) reduces to:

$$\frac{\Delta P_{TP}}{\Delta P_{LO}} = 1 + (\Gamma^2 - 1)[Bx_G(1 - x_G) + x_G^2] \quad (25)$$

Grant and Chisholm[25] report the following relationships for the constant B for window zones:

$$B = \left[\frac{\rho_{hom}}{\rho_L}\right]^{0 \cdot 25} \quad \text{vertical up-and-down flow} \quad (26)$$

$$B = \frac{2}{\Gamma + 1} \quad \text{horizontal side-to-side flow} \quad (27)$$

Again, all of these recommendations are based upon experiments conducted with air/water mixtures.

Most of the studies of tube bundle performance have been restricted to ideal tube bundles (an ideal bundle being one in which edge effects are minimised). Tube bundles that are fitted in shell-and-tube reboilers differ markedly from such arrangements. The construction of these exchangers necessitates clearances between the tube bundle and the shell, and between the tubes and the baffle plates. The effects of these clearances may be significant. Grant *et al.*[26] report the results of experiments with ideal bundles and with bundles containing by-pass channels. With the ideal bundle the quality distribution was found to be practically uniform. However, with the by-pass channel present, it was found that there was a marked maldistribution of liquid and gas with a large liquid flow in the by-pass. A three-fold reduction in pressure drop was observed. Polley and Grant[27] have recently presented a simple model which predicts the effect of by-pass lanes. This model assumes that the pressure drop through the main bundle and through the by-pass lane are equal and that the two phases are distributed in such a way that the pressure loss is a minimum. This model correlated the non-ideal bundle results of Grant *et al.*[26] to an accuracy of ±20% over their full experimental range. This model also provides an estimation of flow distribution through such bundles and although the

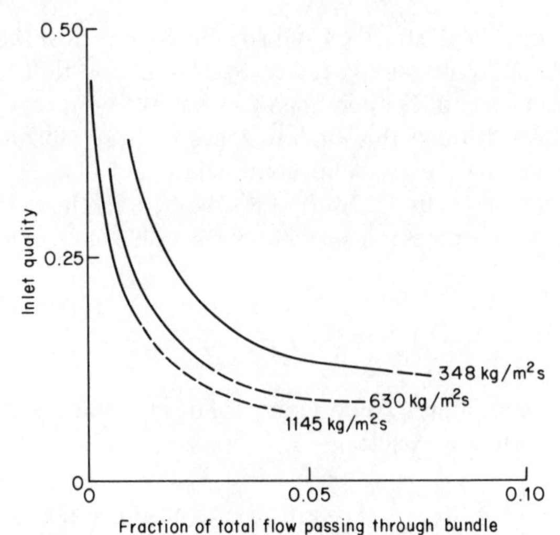

FIG. 3.11. Bundle by-passing with two-phase flow.

predictions of the model have still to be verified by experimental data, they do demonstrate that the boiler designer must take care to reduce by-pass areas to the minimum. The presence of a by-pass lane (in this example equivalent to 20% of the total flow area) can result in the flow of liquid through the bundle being reduced by 75% or more (Fig. 3.11).

Polley *et al.*[28] have investigated flow boiling in tube bundles. They found that two-phase forced flow had a marked effect upon heat transfer with the

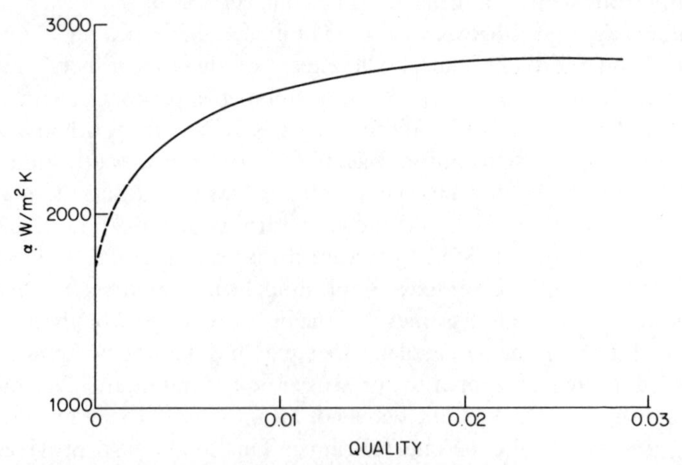

FIG. 3.12. Influence of mass quality during flow boiling in tube bundles.

coefficient increasing markedly with increasing quality of the two-phase mixture (Fig. 3.12). The following equation was proposed for the boiling heat transfer coefficient:

$$\alpha = \alpha_{PB} + \alpha_{FC} \tag{28}$$

where α_{PB} is the pool boiling coefficient calculated at the local superheat and α_{FC} is the two-phase forced convective coefficient given by:

$$\alpha_{FC} = \alpha_L \left[\frac{1}{1 - \varepsilon_g} \right]^{0{\cdot}744} \tag{29}$$

The single-phase heat transfer coefficient α_L for convection in tube bundles can be calculated from relationships proposed by ESDU:[36]

$$\mathrm{Nu} = 0{\cdot}212\,\mathrm{Re}_L^{0{\cdot}651}\,\mathrm{Pr}_L^{0{\cdot}034} \left[\frac{\mathrm{Pr}_L}{\mathrm{Pr}_w} \right]^{0{\cdot}26} \tag{30}$$

The void fraction ε_g, for use in eqn. (29) is estimated from eqn. (23).

4. DESIGN OF KETTLE REBOILERS

How can the theories, correlations and equations described above be used for the design of kettle reboilers? Published design procedures[15,29,30] for kettle reboilers generally assume that vaporisation in tube bundles is predominantly through nucleate boiling. The shell-side heat transfer coefficient is usually based upon a convenient pool boiling correlation (with or without some empirical correction to account for the effects of two-phase forced convection within the bundle). However, using these methods the estimation of the heat transfer coefficient is not very critical in many cases because the maximum permissible heat fluxes recommended by these methods are very conservative.

Kern[29] recommends the following limits:

Boiling coefficient (organics)	300 Btu/h . ft^2 (°F)	(1·7 kW/m^2 (°K))
Boiling coefficient (aqueous)	1 000 Btu/h . ft^2 (°F)	(5·7 kW/m^2 (°K))
Maximum heat flux (organics)	12 000 Btu/h . ft^2	(38 kW/m^2)
Maximum heat flux (aqueous)	30 000 Btu/h . ft^2	(95 kW/m^2)

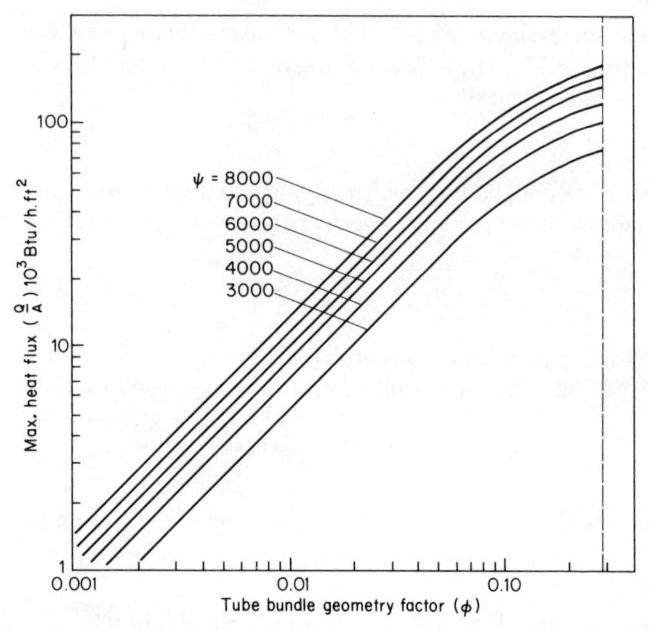

FIG. 3.13. Kettle reboiler limiting heat flux—Palen and Small method.

Frank[31] recommends that the limits on heat flux be raised to:

Maximum heat flux (organics)	20 000 Btu/h . ft^2	(63 kW/m^2)
Maximum heat flux (aqueous)	35 000 Btu/h . ft^2	(110 kW/m^2)

Palen and Taborek[32] obtained performance data from 17 reboilers in operation on refinery plant. (Since these measurements were taken on operational plant it must be assumed that the data refer to the normal operational condition of the reboilers and do not constitute data on actual limiting heat flux.) Many of the observed exchangers were operating at heat fluxes in excess of that recommended by Kern.[29]

In a later paper, Palen and Small[30] present a design guide for limiting heat flux that is based upon these observations. This guide attempts to relate maximum heat flux to bundle geometry (Fig. 3.13) using a tube bundle geometry factor defined as:

$$\phi = 0{\cdot}359\left(\frac{P_t}{D_B}\right)\sqrt{\frac{\sin\alpha}{N}} \tag{31}$$

and a physical property factor defined as:

$$\psi = \rho_v \Delta h_v \left[\frac{g\sigma(\rho_L - \rho_v)}{\rho_v^2} \right]^{0.25} \tag{32}$$

It should be noted that this equation for physical property factor is dimensional. The guide has been produced using British units and is not applicable with any other system of units.

The authors point out that this guide is based upon experience and empirical correlation rather than upon any theory. This guide predicts limiting heat fluxes that are higher than those recommended by Kern, but the comments of Frank[31] indicate that the predictions of this method are still conservative.

Palen *et al.*[15] report in general terms the results of a major study into kettle reboiler performance (the detailed results of the study are proprietary). Their observations regarding heat transfer coefficients are consistent with those discussed above. They suggest that one approach to bundle performance analysis may be to conduct a force balance of two-phase friction, acceleration and static losses in the bundle against the static head of clear liquid surrounding the bundle. However, they did not consider that enough was known about two-phase flow through tube bundles to permit such an approach.

The heat transfer distribution through kettle reboiler bundles reported by Leong and Cornwell (Fig. 3.2) suggest that in addition to a variation in local quality between the bottom and the top of the bundle there is also a variation in quality between the edge and the centre of the bundle. This would suggest that a two-dimensional approach is required if a full analysis of bundle performance is to be achieved. However, the need for such a full analysis must be questioned in the light of the large degree of uncertainty surrounding the prediction of nucleate boiling heat transfer coefficients.

The studies of two-phase flow through tube bundles reported above now make the one-dimensional force balance approach suggested by Palen *et al.*[15] feasible. Brisbane *et al.*[3] have conducted such a study, and that work forms the basis of an effective procedure for kettle reboiler design. This work has demonstrated that the limiting heat flux is controlled by recirculation through the bundle. The recirculation rate is estimated from the force balance. The limiting heat flux is that which results in a recirculation rate that gives an exit quality approaching unity. Typical limiting heat fluxes computed from this model are compared with those suggested by Palen and Taborek[32] in Fig. 3.13. Once the recirculation rate is known, the quality and heat transfer distribution can be estimated using the equations described above.

The kettle reboiler design procedure is as follows:

1. Specify maximum and minimum tube lengths and tube-length interval to be considered for alternative designs. Specify tube diameters and tube-sheet layout.
2. Set tube length at maximum.
3. Estimate approximate heat transfer area from typical film heat transfer coefficients:[33]

	Btu/h ft^2 (°F)	W/m^2 K
Hot side		
Condensing steam	2 000	11 350
Cooling hot water	400	2 270
Cooling hot oil	125	7 100
Boiling side		
C_2—C_4 hydrocarbons	333	1 890
Gasoline and naphtha	200	1 136
Aromatics	333	1 890
C_2—C_7 alcohols	333	1 890
Chlorinated hydrocarbons	250	1 420
Water (atmospheric pressure)	667	3 785

4. Estimate heat flux required for heat duty with this bundle area. Estimate tube count and bundle diameter.
5. Estimate recirculation-limited heat flux for this bundle diameter.
6. If required heat flux estimated in Step 4 exceeds the recirculation-limited heat flux, reduce bundle diameter and return to Step 4. If repetition of this step fails to achieve a suitable bundle, the maximum tube length is too small.
7. Estimate recirculation through the bundle and the mean bundle coefficient. Estimate overall coefficient and compute heat flux.
8. Compare heat flux estimated in Step 7 with that estimated in Step 4. Modify bundle diameter and repeat Steps 5–8 until convergence is achieved.
9. Cost design.
10. Reduce tube length and repeat Steps 3–9 until minimum tube length has been examined.
11. Compare successful designs and select the best.

Obviously this procedure can only be effectively adopted by computer.

5. DESIGN OF THERMOSIPHON REBOILERS

The performance of thermosiphon reboilers is affected by the pipework connecting the reboiler and the liquid reservoir, and the pressure-loss characteristics of this pipework must be known before a design can be achieved. The procedure described below is similar to that proposed by Fair.[33] The calculations involved are long and tedious and can only effectively be conducted by computer. Proprietary computer programs for this task do exist.[34]

The procedure proposed for the design of a vertical thermosiphon reboiler is as follows:

1. Specify pressure-loss characteristics and layout of pipework connecting reboiler and reservoir.
2. Specify maximum and minimum tube lengths and length interval to be considered for alternative designs. (With all designs the level of the liquid in the reservoir is assumed to be that of the top tube-sheet of the reboiler.)
3. Set tube length at maximum.
4. Estimate approximate overall coefficient using typical values in the first instance, but subsequently using the experience of previous calculations.
5. From heat duty, approximate overall coefficient and temperature driving force, estimate required heat transfer area and hence tube count.
6. Guess recirculation rate.
7. Moving step by step along the tube length, estimate heat transfer coefficients in subcooled convective, subcooled nucleate, saturated nucleate and saturated convective regions. Estimate overall pressure drop (including pipe losses) and heat transferred. Variation in saturation temperature along the tube length should be accounted for.
8. Compare overall pressure drop with driving force (liquid static head). If not in balance adjust recirculation rate and repeat Steps 7 and 8.
9. Compare heat transferred with specified duty. If not in agreement adjust tube count and repeat Steps 6–9.
10. Cost design.
11. Adjust tube length and repeat Steps 4–11 until all alternatives have been considered.

In addition to looking at exchanger costs during the evaluation of the alternative designs the engineer should check that the outlet qualities are not excessive and should look at the heat transfer regime distribution. Designs having substantial subcooled zones should be avoided if at all possible.

Where aqueous substances are being evaporated, some live steam injection at the base of the tube bundle may be tolerated. Such injection promotes better heat transfer in the subcooled and low-quality regions of the exchanger. However, great care must be taken to ensure proper flow distribution avoiding preferential flow of steam down some tubes in the bundle.

Alternative designs should be checked for flow stability. Design procedures for horizontal thermosiphon reboilers have not been developed to anywhere near the sophistication that the procedures for kettle and vertical thermosiphon reboilers have reached. The author is not aware of the availability of any computer program for this task, and is only aware of one publication[35] that is specific to horizontal thermosiphon reboiler design. However the principles behind the VTR design procedure also apply to horizontal units and the equations for heat transfer and pressure drop during two-phase flow through tube bundles have been described above. The results of the tube bundle studies suggest two principles:

1. Sealing strips should be employed in horizontal thermosiphon reboilers in order to reduce preferential by-passing and phase separation.
2. The designer should aim to promote a bubbly flow within the reboiler.

What follows is the author's suggestion for a computerised design procedure:

1. Specify pressure-loss characteristics of pipework, etc.
2. Specify maximum tube lengths, etc., as in procedures described above.
3. Start at maximum tube length.
4. Estimate approximate overall coefficient and hence tube count and bundle diameter.
5. Using an approximate recirculation rate based upon a reasonable outlet quality, estimate a baffle spacing profile that will give bubbly flow conditions throughout the exchanger.
6. Estimate recirculation rate.

7. Conduct a baffle-space-by-baffle-space analysis of heat transfer performance and pressure drop. Estimate overall performance.
8. Compare pressure loss with driving force. Adjust recirculation rate and repeat Steps 6 and 7 until agreement is achieved.
9. Compare heat transfer performance and duty. Adjust tube count if necessary and repeat Steps 6–9 until agreement is achieved.
10. Cost design; then adjust tube length and examine alternatives.

Such a procedure will inform the designer of probable cost-effective design configurations. The baffle arrangements of such designs must then be refined in order to obtain bubbly-flow conditions throughout the exchanger. Experience in the use of the above procedure will lead to refinement in the method of deriving an initial baffle arrangement (Step 5).

6. DESIGN OF FORCED CIRCULATION REBOILERS

The design of forced circulation reboilers is fairly straightforward and can be summarised as follows:

1. Set recirculation rate (based upon reasonable outlet quality), allowable pressure drop, maximum and minimum tube lengths, etc., and baffle spacing profile if relevant (shell-side boiling).
2. Set tube length and conduct approximate calculation to obtain starting point for tube count estimation.
3. Conduct step-by-step (or baffle-space-by-baffle-space) analysis of heat transfer and pressure drop performance.
4. Compare predicted performance with required duty and adjust tube length if necessary (repeating Steps 3 and 4).
5. Compare predicted pressure drop with allowable pressure drop. Flag unsuitable designs.
6. Cost design.
7. Reduce tube length and repeat Steps 2–7.

The designer has a number of process parameters to play with: recirculation rate, allowable pressure drop (even in some cases heat load). These should be selected on the basis of minimum annual operating cost.

As in every aspect of engineering there is a trap for the unwary. As flow through a boiler is varied, the variation of pressure drop with flow may not be progressively increasing. It could follow the direction illustrated in Fig. 3.14. Provided the pump curve only crosses this curve at one point, a stable

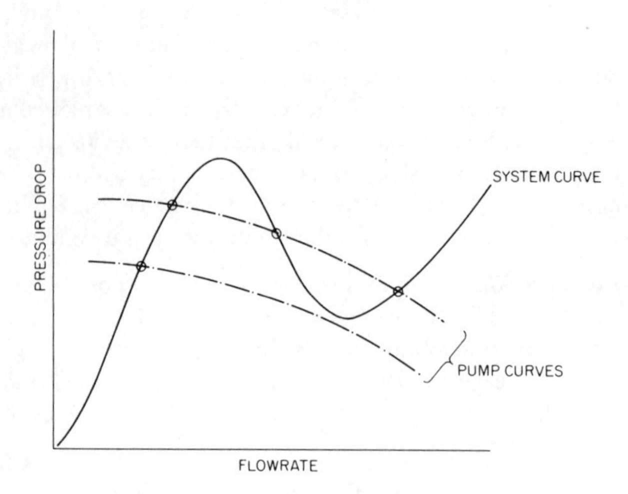

FIG. 3.14. Interaction between two-phase pressure drop/flow characteristic and pump curve.

calculable operation-point exists. If the pump curve cuts this curve in more than one point an excursive instability can occur and the operational point may be other than that expected from the initial design calculations.

REFERENCES

1. *Standards of the Tubular Exchanger Manufacturers Association* (*TEMA*), 6th ed., 1979, Tubular Exchanger Manufacturers Association, New York.
2. LEONG, L. S. and CORNWELL, K. Flow boiling heat transfer coefficients in a kettle reboiler tube bundle, *The Chemical Engineer*, 1979, **57**, 219.
3. BRISBANE, T. W. C., GRANT, I. D. R. and WHALLEY, P. B. A prediction method for kettle reboiler performance, Paper presented at 19th National Heat Transfer Conference ASME/AIChE, July 1980, Orlando, Fl.
4. LORENZ, J. J., MIKIC, B. B. and ROHSENOW, W. M. The effect of surface conditions on boiling characteristics, *Heat Trans.*, 1974, **4,** 35–39.
5. FUJII, M., NISHIYAMA, E. and YAMANAKA, G. Nucleate pool boiling heat transfer from micro-porous heating surface, Paper presented at 18th National Heat Transfer Conference ASME/AIChE, July 1979, San Diego, Calif.
6. MILTON, R. M. and GOTTZMAN, G. F. High efficiency reboilers and condensers, *Chem. Eng. Prog.*, 1972, **68**(9), 56–61.
7. BAILEY, N. A. Introduction to hydrodynamic instability, Chapter 16 in: *Two Phase Flow and Heat Transfer*, D. Butterworth and G. F. Hewitt (eds.), 1977, Oxford University Press, London.
8. POLLEY, G. T. Correlations for forced convection heat transfer, *The Chemical Engineer*, 1979, **57,** 233–34.

9. Davis, E. J. and Anderson, G. H. The incipience of nucleate boiling in forced convection flow, *AICHE J.*, 1966, **12**(4), 774–80.
10. Chen, J. C. Correlations for boiling heat transfer to saturated fluids in convective flow, *I.E.C. Process Des. Dev.*, 1966, **5**(3), 322–29.
11. Moles, F. D. and Shaw, J. F. G. Boiling heat transfer to sub-cooled liquids under conditions of forced convection, *Trans. Inst. Chem. Eng.*, 1972, **50,** 76–84.
12. Bergles, A. E. and Rohsenow, W. M. The determination of forced convection surface boiling heat transfer, *J. Heat Trans.*, 1964, **86,** 365–72.
13. Dougall, R. S. and Panian, D. J. Subcooled forced convection boiling of trichlorotrifluoroethane, NASA CR-2137, 1972.
14. Voloshko, A. A. Free convection boiling of Freons, *Heat Trans.-Sov. Res.*, 1972, **4**(4), 60–66.
15. Palen, J. W., Yarden, A. and Taborek, J. Characteristics of boiling outside large scale horizontal multi-tube bundles, *AIChE Symp. Ser.*, 1972, **68**(118), 50–61.
16. Mostinski, I. L. Applications of the law of corresponding states to the calculation of heat transmission to boiling fluids and of their critical heat flux, *Teploenergetika*, 1963, **10**(4), 66.
17. Hsu, Y. Y. and Graham, R. N. *Transport Processes in Boiling and Two Phase Systems*, 1976, Hemisphere Publishing Corp., Washington, D.C.
18. Guerrieri, S. A. and Talty, R. D. A study of heat transmission to organic liquids in single tube, natural circulation, vertical tube boilers, *Chem. Eng. Prog. Symp. Ser.*, 1956, **52**(18), 69–77.
19. Collier, J. G. *Convective Boiling and Condensation*. 1972, McGraw-Hill, London.
20. Dukler, A. E., Wicks, M. and Cleveland, R. G. Frictional pressure drop in two phase flow: a comparison of existing correlations for pressure loss and hold-up, *AIChE J.*, 1964, **10,** 38–43.
21. Chisholm, D. A theoretical basis for the Lockhart–Martinelli correlation for two-phase flow, *Int. J. Heat Mass Trans.*, 1967, **10,** 1767–78.
22. Lockhart, R. W. and Martinelli, R. C. Proposed correlation of data for isothermal two-phase two-component flow in pipes, *Chem. Eng. Prog.*, 1949, **45,** 39–48.
23. Chisholm, D. Pressure gradients due to friction during the flow of evaporating two-phase mixtures in smooth tubes and channels, *Int. J. Heat Mass Trans.*, 1973, **16,** 347–58.
24. Chisholm, D. Void fraction during two-phase flow, *J. Mech. Eng. Sci.*, 1973, **15**(3), 235–36.
25. Grant, I. D. R. and Chisholm, D. Two phase flow on the shell-side of a segmentally baffled shell-and-tube heat exchanger, *J. Heat Trans.*, 1979, **101,** 38–42.
26. Grant, I. D. R., Finlay, I. C. and Harris, D. Flow and pressure drop during vertically upward two-phase flow past a tube bundle with and without bypass leakage, *Ind. Chem. Eng. Symp. Ser.*, 1974, **2**(38), paper I1.
27. Polley, G. T. and Grant, I. D. R. Pressure drop prediction for two-phase upward flow through a tube bundle with bypassing, Paper A10, *Proc. Int. Meeting on Industrial Heat Exchangers, Liege, Nov. 1979*, Assoc. Eng. Grad. Univ. Liege.

28. POLLEY, G. T., RALSTON, T. and GRANT, I. D. R. Forced crossflow boiling in an ideal in-line tube bundle, Paper presented at 19th National Heat Transfer Conference ASME/AIChE, July 1980, Orlando, Fl.
29. KERN, D. Q. *Process Heat Transfer*, 1950, McGraw-Hill, New York.
30. PALEN, J. W. and SMALL, W. M. A new way to design kettle and internal reboilers, *Hydrocarbon Processing*, 1964, **43**(11), 199–208.
31. FRANK, O. Simplified design procedures for tubular heat exchangers, Chapter 1 in: *Practical Aspects of Heat Transfer*, 1978, AIChE, New York.
32. PALEN, J. W. and TABOREK, J. Refinery kettle reboilers, *Chem. Eng. Prog.*, 1962, **58**(7), 37–46.
33. FAIR, J. R. What you need to design thermo-siphon reboilers, *Petroleum Refiner*, 1960, **39**(2), 105–23.
34. PETERSON, J. N., CHAU-CHYUN, C. and EVANS, L. B. Computer programs for chemical engineers, 1978—Part 2, *Chem. Eng.*, 1978, July 3, 69–82.
35. COLLINS, G. K. Horizontal thermosiphon reboiler design, *Chem. Eng.*, 1976, July 19, 149–52.
36. ENGINEERING SCIENCES DATA UNIT. Convective heat transfer during crossflow of fluids over plain tube banks, 1973, Item No. 73031, Engineering Sciences Data Unit, 251–59 Regent St., London.
37. STEPHAN, K. and ABDELSALAM, M. Heat transfer correlations for natural convection boiling, *Int. J. Heat Mass Trans.*, 1980, **23**, 73–87.

Chapter 4

CONDENSERS

D. R. WEBB
University of Manchester,
Institute of Science and Technology,
Manchester, UK

and

J. M. McNAUGHT
National Engineering Laboratory, East Kilbride,
Glasgow, UK

SUMMARY

The chapter on condensers provides a fairly detailed description of the process of condensation and the factors which are important in design. The emphasis is on thermal design and consideration is limited to those methods which might reasonably be applied in industrial practice and their comparison. These are methods based on the assumption of local vapour–liquid equilibrium, the equilibrium methods, and the so-called 'film model'. Particular emphasis is placed on recent developments in the understanding of the process of multi-component mass transfer and their impact on condenser design. Methods for the incorporation of such ideas into robust design procedures are introduced. A feature of the presentation is the inclusion of comprehensive examples dealing with simple condenser geometries which should facilitate the implementation and subsequent verification of the design procedures on a digital computer. The thermal and pressure-drop design of condensers of more complex geometry is discussed and a comprehensive list of references is given to sources giving additional information.

NOTATION

$[B]$	matrix of multicomponent mass transfer coefficients
c_p	specific heat capacity
d	tube inside diameter
h	specific enthalpy
Δh_v	latent heat of vaporisation
j	molar diffusive flux density (kmol/m^2 s)
$\dot{M}$	mass flowrate (kg/s)
$\tilde{M}$	molecular weight
$\dot{N}$	molar flowrate (kmol/s)
$\dot{n}$	molar total flux density (kmol/m^2 s)
P	pressure
Pr	Prandtl number
R	universal gas constant
Re	Reynolds number
S	tube cross-sectional area
Sc_{ij}	Schmidt number associated with binary diffusivity δ_{ij}
x	vapour mass flow fraction (quality)
α	heat transfer coefficient
β_{ij}	binary mass transfer coefficient (kmol/m^2 s)
λ	thermal conductivity
μ	dynamic viscosity
$\tilde{\theta}$	vapour fraction on molar basis
$[\Phi]$	matrix of multicomponent rate factors
ρ	density

Superscripts

*	equilibrium quantity
~	molar quantity
· (as in $\alpha^{\cdot}$)	modified heat or mass transfer coefficient

Subscripts

c	coolant
f	gas film
g	gas phase
go	total flow as gas
I	interface
l	liquid phase
lo	total flow as liquid

m	mean
s	sensible (heat)
T	total
TP	two-phase
z	zone

1. INTRODUCTION

1.1. Condensation in the Power and Process Industries

The condenser is widely found both in the power generation and chemical process industries. In the power generation field the condenser is the final heat sink in the thermodynamic cycle, while in the process industries it is most commonly used to condense the overheads from distillation columns. However, condensation also occurs in many other types of heat exchanging equipment, such as feed heaters and gas coolers.

The condenser plays an extremely important role in the power industry. Maximum efficiency results when heat is rejected in the condenser at the lowest possible temperature. Consequently the condenser must be designed to operate with a small temperature difference between the condensing medium and the coolant, and steam power condensing plant must therefore be capable of producing and maintaining vacuum conditions. In condensing the overheads of a distillation column the operating pressure determines the condensing temperature. The condenser must therefore be capable of handling the required load with the temperature difference available between the saturation temperature corresponding to the operating pressure and the coolant temperature. The designer is restricted in this task by the allowable pressure drop, which is generally rather low. Failure to accomplish a successful design will cause a restriction in the column throughput.

The basic hardware of condensing plant is dictated by the available space, economics and the conditions of the process, which generally require segregation of the coolant and the condensing stream, and a low vapour pressure drop. In the power generation industry, where the coolant is often seawater or river water, and frequent cleaning of the surface exposed to the coolant is required, it is convenient to pass the coolant through smooth tubes. The design which has evolved is known as the surface condenser, where steam is condensed on the outside of horizontal bundles of tubes through which flows cooling water. In the process industries the baffled shell-and-tube heat exchanger is commonly used for condensing duties.

Condensation can take place inside or outside the tubes, which are usually positioned either horizontally or vertically.

Much less commonly used than the surface and shell-and-tube condenser is the direct contact condenser, where a vapour is condensed either by bubbling it through a cooling medium, or by bringing it into contact with a spray of coolant. Application of the direct contact condenser is restricted to cases in which it is permissible to mix the vapour and coolant, or where the coolant and condensable can be easily separated.

No review of developments in condenser technology would be complete without mentioning dropwise condensation. Heat transfer coefficients in the dropwise mode can be an order of magnitude greater than in film condensation, but the practical problems of guaranteeing and maintaining dropwise condensation in industrial plant remain unsolved. A recent review by Tanasawa[1] also indicates that much remains to be learned about the basic mechanism of dropwise condensation. He suggests that polymer coatings are the most promising method of maintaining dropwise condensation in practical applications.

An alternative means of enhancing the heat transfer coefficient, by means of specially shaped tubes, is discussed elsewhere in this book.

1.2. Scope of this Work

While providing a reasonably detailed description of the process of condensation and the factors which are important in condenser design, this chapter emphasises recent developments in the understanding of multicomponent mass transfer and their impact on condenser design. Methods for the incorporation of such ideas into robust and reliable design procedures for condensers are described. The more common design methods used in industrial practice are also covered for completeness and for comparison purposes. A feature of the presentation is the inclusion of comprehensive examples, to facilitate the implementation and subsequent verification of the design procedures on a digital computer.

2. THE CONDENSATION PROCESS

Condensation is the process whereby a vapour is reduced to the liquid state by removal of its latent heat of vaporisation. The process may occur in a number of different ways depending on local conditions and the geometry of the condenser.

2.1. Modes of Condensation

When a pure vapour, vapour mixture or vapour containing a non-condensable gas is brought into contact with a surface below its saturation temperature or dew point, condensation occurs. The condensate may form a continuous film over the cold surface or discrete droplets, depending on whether the surface is wetted or not. Filmwise condensation is normally encountered in industrial applications and dropwise condensation can only be maintained under controlled conditions with special surface coatings or additives to the vapour. Where the vapours are of immiscible liquids the condensate structure and flow pattern are more complex. The liquid phase which preferentially wets the surface tends to form a film while the second liquid may form standing drops, lenses floating on the surface of the surface-wetting liquid or liquid rivulets. If the vapour becomes super-saturated it may undergo homogeneous condensation or fogging, though the presence of liquid nuclei or foreign particles appears to be essential for fog formation.

Filmwise condensation leads to far higher resistances to heat transfer than dropwise condensation and is usually designed for. In partial condensation it is important to be able to check for conditions favouring fog formation since its occurrence leads to problems of entrainment and loss of condensable product.

Condensation may also be induced by contacting the vapour with a liquid at a temperature below its dew point, known as direct contact condensation. The liquid may be sprayed into the vapour or the vapour injected into a pool of liquid in single-stage processes, or countercurrent contactors such as packed columns may be used.

2.2. Condenser Geometry

The cooled surface in practical condensers may be of any orientation, though vertical and near-horizontal geometries are preferred. In shell-and-tube exchangers the condensing vapours may be fed to either shell- or tube-side depending on the nature of the fluids involved, their pressure, tendency to cause corrosion and fouling characteristics. Consideration must therefore be given to a wide range of flow arrangements.

In vertical condensers with condensation inside or outside tubes, a film of condensate will fall under the influence of gravity thickening downstream with increasing load. Film flow will be laminar near the top of the tube but may become turbulent at high loadings (Fig. 4.1(a)). The film surface is usually covered with ripples or waves which influence the condensation process. The influence of the vapour flow is important. If the vapour flow is concurrent with condensate, the surface shear causes film thinning. On the other hand a small countercurrent flow will cause film thickening (Fig. 4.1(b),(c)) while a larger flow may cause flooding and eventually flow reversal (Fig. 4.1(d),(e)). A flow of vapour across tubes will lead to a non-axisymmetric liquid distribution with thinning of the film on the upstream side and thickening in the wake.

The heat transfer resistance of the condensate film is directly proportional to its thickness and is reduced by wave effects and turbulence.

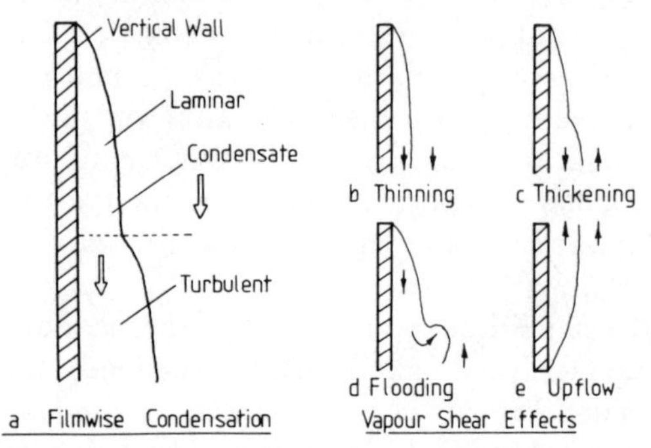

FIG. 4.1. Film condensation on vertical surfaces.

Cocurrent and cross-vapour flows cause film thinning and reduce heat transfer resistance. On the other hand countercurrent flows should be avoided. Conditions well below flooding are permissible but it is not recommended that design should be carried out for the case of reverse flows (Fig. 4.1(e)).

In condensation outside a bank of horizontal tubes the vapour may flow vertically upwards or downwards across the tubes, or horizontally in a direction parallel to or perpendicular to the tubes. Usually horizontal condensers are vertically baffled to produce a combination of horizontal cross flow and parallel flow. Vapour upflow is not normally recommended because of the possibility of flooding. In perfectly horizontal tubes the condensate drips vertically from the uppermost tubes leading to thicker films and higher heat transfer resistance of the lower tube rows, known as condensate inundation. If the tubes are inclined at even modest angles to the horizontal, the liquid flows in the direction of the slope at least as far as any vertical baffle (Fig. 4.2).

Condensation may also be arranged to occur within horizontal tubes, the flow characteristics depending on the vapour velocity. At low vapour flow-rates gravity effects are dominant. A film of condensate of variable

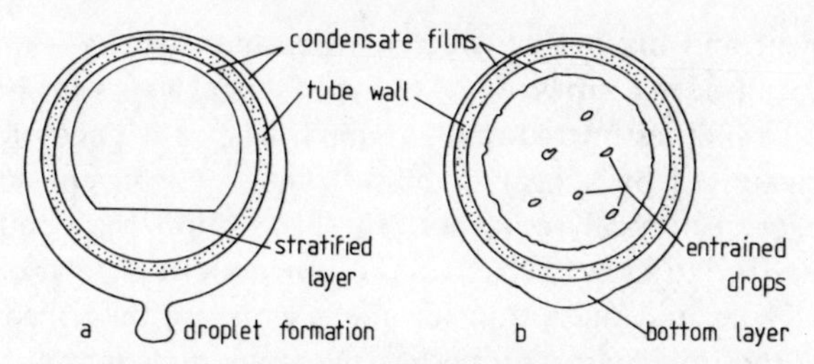

FIG. 4.2. Condensation inside and outside horizontal tubes. (a) Low inside and outside velocity, horizontal inclination. (b) High inside velocity, low outside velocity, moderate inclination.

thickness forms around the tube and drains into a stratified layer at the bottom. This layer passes from the tube under its own hydraulic gradient or with the slope. At high vapour flowrates vapour shear effects, including the formation and deposition of entrained droplets give rise to films which are almost axisymmetric (Fig. 4.2).

As with condensation on vertical surfaces, the effect of vapour shear is to cause film thinning and lower resistance to heat transfer. However, at low vapour velocities, liquid film flow is usually laminar because of the smaller local loadings.

In condensation in horizontal condensers it is clear that the vapour may become separated from the condensate if not as a result of condensate inundation then through the effect of vertical baffles. In vertical condensers the same effect is produced if the film of condensate is not well mixed. This may well have a serious effect on the condensation of multicomponent vapours where less volatile components will condense preferentially near the vapour inlet. The bubble point of the mixture of the lighter components remaining in the vapour is decreased and may fall below the cooling fluid temperature preventing the total condensation of a mixture, theoretically possible without flow separation. Design is inherently more difficult where flow separation occurs because the problem becomes two-dimensional.

3. THE MECHANISM OF CONDENSATION

Reliable design methods for condensers are based on a clear understanding of the underlying processes of heat and mass transfer. The actual physical situation is extremely complex and, at the present state of development, practical design methods are at their most sophisticated based on the film model, a somewhat crude approximation to the physical reality. The transfer of heat and mass is considered to be impeded by a number of resistances, localised in a series of real or hypothetical layers or films. These must be individually estimated and summed. The resistance of a film is usually characterised by a heat or mass transfer coefficient which is a conductance or reciprocal resistance. In this section the condensation process is described to show the basis of the film model. Correlations for the evaluation of heat and mass transfer coefficients are described and the transport and thermodynamic properties, necessary in their application are listed.

3.1. The Resistances to Heat and Mass Transfer in Condensation—Film Coefficients

Consider the process of condensation brought about by contacting a pure vapour, a vapour–gas mixture or a vapour–gas mixture, saturated or superheated, and of temperature T_g, with a cold surface below its saturation temperature. The latent and sensible heats are carried away by a cooling medium of temperature T_c, which maintains the surface temperature of the wall T_w. The composition of the vapour phase by mole fraction is $\tilde{y}_{ig}$ where subscript i is the component number in a mixture of n components. The quantities T_c, T_g and $\tilde{y}_{ig}$ of which $(n + 1)$ are independent in a mixture of n components are assumed to be known locally and the local design problem is the calculation of the local rates of heat and mass transfer.

The transfer of heat through the wall and a hypothetical stagnant film of coolant is by conduction alone, and in this work it will be assumed that a heat transfer coefficient, α_c, may be defined so that the energy flux, $\dot{q}_1$, is given by

$$\dot{q}_1 = \alpha_c(T_w - T_c) \tag{1}$$

The coefficient α_c, reflects the resistances to heat transfer of coolant film (for which correlations are available for a wide range of geometries), the wall, and dirt or fouling layers which may be present or allowed for.

There are as many as three resistances to heat and mass transfer which may be considered to arise on the vapour side of the wall: the condensate resistance, the interfacial resistance and the resistance of the vapour film.

A condensate resistance is always present in condensation, though it may arise from the presence of a continuous film, droplets or a combination of these. The heat flux in the condensate film (which may be real or hypothetical) will vary and the temperature profile (Fig. 4.3) will be non-linear due to subcooling of the condensate. However, the sensible heat of subcooling is always small compared to the latent heat and may be neglected or taken into account by assuming that $\dot{q}_l$, corrected for subcooling, is constant across the liquid film. Methods of predicting the heat transfer coefficient of the liquid film α_l, which account for laminar or turbulent film flow, vapour shear and wave effects, are discussed in Section (3.3).

Where the condensate contains more than one component, liquid-side mass transfer effects may become important, but there is little published

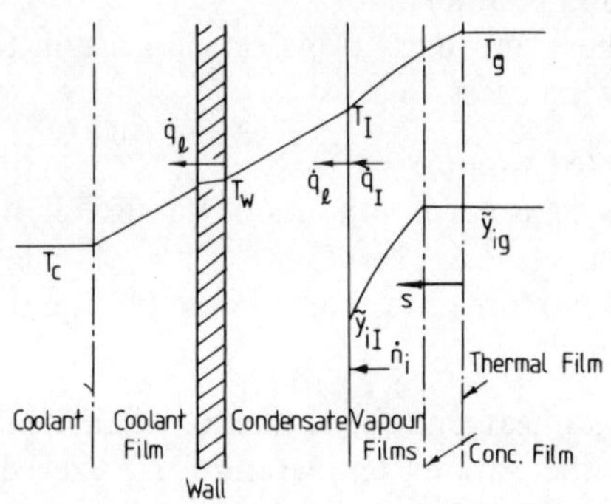

FIG. 4.3. Film model of condensation.

information on this topic. Two limiting cases may be described. The first is of complete liquid mixing where the local liquid composition $\tilde{x}_{il}$, is determined from the amount of prior condensation. The second is of no liquid mixing and the surface composition is determined by the ratio of fluxes of condensation $\dot{n}_i$, to the total flux $\dot{n}_T$.

$$\tilde{x}_{il} = \dot{n}_i/\dot{n}_T \tag{2}$$

The case of no liquid mixing is applicable where the condensate is continuously separated from the vapour (see Section (4.2)), a feature of horizontal tube condensation. There is evidence in the work of Deo[2] and Sardesai[3] that liquid mixing is not important where the vapour contains a high concentration of non-condensable gas.

In the absence of information on liquid-side mass transfer, the interfacial temperature T_I, is given by

$$\dot{q}_l = \alpha_l(T_I - T_W) \tag{3}$$

There is evidence,[4] based on the kinetic theory of gases, that an interfacial resistance may arise in condensation. However, the experimental measurement of such effects is difficult, though for single vapours an effective heat transfer coefficient may be defined and predicted according to the methods of Section (3.3). There is no such information available on multicomponent mixtures, and it is necessary to assume vapour–liquid equilibrium at the interfacial temperature T_I, to provide the relationship between the interfacial vapour and liquid compositions $\tilde{y}_{iI}$ and $\tilde{x}_{iI}$.

$$\tilde{y}_{iI} = K_i \tilde{x}_{iI} \tag{4}$$

The K values will in general be a function of temperature T_I, pressure P, and the vapour and liquid compositions.

A resistance to heat and mass transfer will arise in the vapour phase except in the following cases:

(i) Pure saturated vapours.
(ii) A saturated mixture of vapours of immiscible liquids of eutectic composition.
(iii) A saturated mixture of vapours of miscible liquids at an azeotropic composition.

Even in these cases, a heat transfer resistance will arise if the vapours are superheated when the vapour temperature T_g, exceeds the interfacial temperature T_I (Fig. 4.3). With the heat transfer coefficient α_g, calculated by the methods of Section (3.3), the conductive heat flux $\dot{q}_I$, at the vapour–liquid interface is given by Krishna *et al.* as,[5]

$$\dot{q}_I = \alpha_g^{\cdot} e^{\varepsilon}(T_g - T_I) = \alpha_g e^{\varepsilon}\varepsilon(T_g - T_I)/(e^{\varepsilon} - 1)$$

where,

$$\varepsilon = \sum \dot{n}_i \tilde{c}_{pi}/\alpha_g = \dot{n}_T \tilde{c}_p/\alpha_g$$

where,

$$\tilde{c}_p = \sum \frac{\dot{n}_i}{\dot{n}_T} \tilde{c}_{pi} \tag{5}$$

The Ackermann correction factor $\varepsilon/(e^{\varepsilon} - 1)$, was included by Ackermann[6] and Colburn and Drew[7] to allow for the transfer of sensible heat by the condensing species.

Mass transfer resistance will arise where the vapour contains two or more components, provided cases (ii) or (iii) (above) are not realised. In general, the heavier components (less volatile) of a vapour mixture have lower diffusivities with respect to the mixture than the more volatile components but at the same time a higher tendency to condense. The result is an accumulation of lighter species near the condensing interface which hinders the arrival of heavier components. This resistance to mass transfer is particularly pronounced where a non-condensable gas is present. The total flux of condensation of a component is made up of a diffusive and convective contribution,

$$\dot{n}_i = j_{ig} + \tilde{y}_{ig}\dot{n}_T \tag{6}$$

In binary mixtures there is only one independent concentration driving force $(\tilde{y}_{ig} - \tilde{y}_{il})$ between bulk and interface, and j_{ig} is obtained from Fick's Law, which with the definition of a mass transfer coefficient β_{gin}, as in Section (3.3), gives

$$j_{ig} = \beta_{gin}\phi_i(\tilde{y}_{ig} - \tilde{y}_{il})/(e^{\phi_i} - 1) = \beta^{\cdot}_{gin}(\tilde{y}_{ig} - \tilde{y}_{il})$$

where, $i = 1, \ldots (n-1)$

$$\phi_i = \dot{n}_T/\beta_{gin} \tag{7}$$

Equation (7) is only strictly true for a binary mixture (with $i = 1, n = 2$) but its predictions are remarkably accurate for many cases of multicomponent condensation. Note the similarity of the correction factors ε and ϕ_i. The strictly accurate multicomponent form of eqn. (7) is presented below.

The closure of the above equations is achieved by the condition of energy continuity at the interface.

$$\dot{q}_l = \dot{q}_I + \sum_{i=1}^{n} \dot{n}_i \Delta\tilde{h}_{vi} \tag{8}$$

The method of solution of the above set of equations will depend on whether the liquid is assumed to be mixed when $\tilde{x}_{il}$ is specified or unmixed when $\tilde{x}_{il}$ must be calculated by eqn. (2). Computer algorithms for the solution of these equations are considered later.

Multicomponent mass transfer may be described by a generalised form of Fick's Law, which is empirical, or in the special case of ideal vapour mixtures under isothermal, isobaric conditions, by the Maxwell–Stefan equations.

The generalised Fick's Law formulation has been applied to a film model of interphase mass transfer in the linearised theories of Toor[8] and Stewart and Prober.[9] More recently Krishna and Standart[10] have developed an exact solution of the Maxwell–Stefan equations for a film model. Differing assumptions are made in the two approaches but the final results are very similar. Given the physical situation of Fig. 4.3 with all vapour films assumed to be of equal thickness, the diffusive flux of a component in an n component mixture is given by

$$(j_g) = [B][\Phi]\{\exp[\Phi] - \ulcorner I \lrcorner\}^{-1}(\tilde{y}_g - \tilde{y}_l) = [B^{\cdot}](\tilde{y}_g - \tilde{y}_l) \tag{9}$$

In this equation (j_g) and $(\tilde{y}_g - \tilde{y}_l)$ are column matrices of dimension $(n-1)$, $[B]$ and $[\Phi]$ are square matrices of dimension $(n-1) \times (n-1)$ and $\ulcorner I \lrcorner$ is the diagonal unit matrix of dimension $(n-1)$. Comparison with eqn. (7) shows the similarity of the full matrix and simplified approaches. In matrix notation, eqn. (7) may be written

$$(j_g) = \ulcorner \beta_{gn} \lrcorner \ulcorner \frac{\phi}{\exp\{\phi\} - 1} \lrcorner (\tilde{y}_g - \tilde{y}_l) \tag{10}$$

From the comparison of eqns. (9) and (10) it is clear that $[B]$ may be interpreted as a matrix of mass transfer coefficients and $[\Phi]\{\exp[\Phi] - \ulcorner I \lrcorner\}^{-1}$ is the multicomponent analogue of the high flux correction of eqn. (10).

Equation (10) is the limiting form of eqn. (9) in the case of equal diffusivities of all binary pairs of components in the mixture, and in the limit of dilute mixtures of all condensable components in a non-condensable gas. In the case of ternary condensation in the presence of a non-condensable gas, it is possible to predict whether eqn. (10) is an adequate representation of the full eqn. (9).[11]

The condensation fluxes $\dot{n}_i$, are constant across a planar film but the diffusive fluxes, j_i, vary. Subscript g indicates that a diffusive flux j_{ig} has been evaluated at the bulk vapour end of the diffusion film, but it may also be evaluated at the interface. The origin, $s = 0$, could equally well have been located at the condensate surface rather than the bulk vapour end as in Fig. 4.3, and this would perhaps have been a more logical choice since the condensate–vapour interface is a real plane. Wherever the origin is located the j_i may in principle be evaluated at either end of the film. However, it is necessary to evaluate j_{ig} to ensure that design using the computational algorithm suggested by Krishna and Standart[10] is convergent.[12]

It is convenient to define a matrix $[M]$ which is related to a column matrix (m) by

$$M_{ij} = \frac{m_i}{\beta_{gin}} + \sum_{\substack{l=1 \\ l \neq i}}^{n} \frac{m_l}{\beta_{gil}}$$

$$M_{ij} = -m_i \left\{ \frac{1}{\beta_{gij}} - \frac{1}{\beta_{gin}} \right\} \qquad i,j = 1, n-1 \tag{11}$$

The following table then shows how $[B]$ and $[\Phi]$ should be evaluated by the matrix theories.

TABLE 4.1

Method	*Matrix* [M]	*Column matrix* (m)
Toor[8]	$[B]^{-1}$	$(m) = (\tilde{y}_g + \tilde{y}_l)/2{\cdot}0$
	$[\Phi]$	$(m) = \dot{n}_T(\tilde{y}_g + \tilde{y}_l)/2{\cdot}0$
Krishna and Standart[10]	$[B]^{-1}$	$(m) = (\tilde{y}_g)$
	$[\Phi]$	$(m) = (\dot{n})$

The binary mass transfer coefficients β_{gij}, are calculated by the methods discussed in Section (3.3).

Equation (9) allows the calculation of the $(n - 1)$ independent diffusive fluxes, j_{ig}. However there are n independent fluxes of condensation $\dot{n}_i$, and determination of the $\dot{n}_i$ requires a determinacy condition. In the computational algorithms presented in Section (4.2) the determinacy condition will be that $\dot{n}_T$, the total flux of condensation is specified. However, another has been specified by Krishna and Standart[13] for the case of condensation in the presence of a non-condensable gas, component n, where $\dot{n}_n$ is taken as zero. It can then be shown that the $\dot{n}_i$ may be determined from the j_{ig} using the expression

$$(\dot{n}) = [A](j_g) \tag{12}$$

where,

$$A_{ij} = \delta_{ij} + \tilde{y}_{ig}/\tilde{y}_{ng} \qquad \begin{matrix} \delta_{ij} = 1 & i = j \\ \delta_{ij} = 0 & i \neq j \end{matrix} \tag{13}$$

However no component is truly non-condensable. Hydrogen for example exhibits a small but finite solubility in a mixture of hydrocarbons. Moreover this determinacy condition will require that the method of calculating the vapour–liquid equilibrium be artificially modified to ensure that the non-condensable gas is not treated in the normal manner.

The equations presented in this section allow the calculation of local rates of heat and mass transfer in a condenser, for binary and multicomponent mixtures. The construction of a computer algorithm for solving these equations is discussed in Section (4.2).

3.2. The Use of Heat Transfer Coefficients in Design

The overall design equation for any heat exchanger allows the estimation of the total duty $\dot{Q}_T$, in terms of the mean overall heat transfer coefficient U_m, the total heat exchanger area A_T, and the mean temperature difference between the hot and cold streams ΔT_m.

$$\dot{Q}_T = U_m A_T \Delta T_m \tag{14}$$

In a differential area dA, where the local overall coefficient is U and the temperature driving force is ΔT, the differential rate of heat transfer is given by

$$d\dot{Q} = U \Delta T dA \tag{15}$$

Integration of this equation over the condenser gives either:

(i) an expression explicit in $\dot{Q}_T$, which may be used for rating an existing unit, or
(ii) an expression explicit in A_T, which is the true design equation for calculation of the area required to achieve a given duty, or
(iii) an expression which in association with eqn. (14) allows the definition of the appropriate mean temperature driving force ΔT_m, and overall heat transfer coefficient U_m.

$$\dot{Q}_T = \int_{A_T} U \Delta T dA \qquad \text{(rating equation)} \tag{16}$$

$$A_T = \int_{\dot{Q}_T} \frac{d\dot{Q}}{U \Delta T} \qquad \text{(design equation)} \tag{17}$$

$$\int_{\dot{Q}_T} \frac{d\dot{Q}}{\Delta T} = \int_{A_T} U dA \tag{18}$$

Comparing eqns. (18) and (14) it follows that

$$\frac{1}{\Delta T_m} = \frac{1}{\dot{Q}_T} \int_{\dot{Q}_T} \frac{d\dot{Q}}{\Delta T} \quad \text{and} \quad U_m = \frac{1}{A_T} \int_{A_T} U dA \tag{19}$$

If ΔT varies linearly with $\dot{Q}$, which occurs when the streams between which heat is to be transferred follow a linear temperature–enthalpy relationship, it follows that

$$\Delta T_m = \Delta T_{ln} = (\Delta T_{in} - \Delta T_{out}) / \ln \left\{ \frac{\Delta T_{in}}{\Delta T_{out}} \right\} \tag{20}$$

This is only satisfied exactly in condensation of pure saturated vapours and saturated azeotropic or eutectic mixtures where the vapour temperature remains constant. Thus the use of an overall log mean temperature driving force is not appropriate in condensation except where the vapour temperature of a saturated vapour remains constant.

Let us consider the condensation of a pure saturated vapour where the local heat transfer coefficient of the condensate film α_l, decreases downstream and the heat transfer coefficient α_c, of the liquid side is assumed constant. Then the local overall heat transfer coefficient U, will be

$$U = \alpha_l \alpha_c / (\alpha_l + \alpha_c) \tag{21}$$

and the appropriate overall mean heat transfer coefficient follows as

$$U_m = \frac{1}{A_T} \int_{A_T} \frac{\alpha_l \alpha_c}{(\alpha_l + \alpha_c)} \mathrm{d}A \tag{22}$$

but,

$$U_m \neq \frac{\bar{\alpha}_l \alpha_c}{\bar{\alpha}_l + \alpha_c} \tag{23}$$

where,

$$\bar{\alpha}_l = \int_{A_T} \alpha_l \, \mathrm{d}A$$

It is the overall mean coefficient $\bar{\alpha}_l$ which is usually quoted for use in condenser design and U_m is then calculated by eqn. (23). This is clearly an approximation, for U_m is correctly evaluated by integration of eqn. (22). In Section (3.3), dealing with the evaluation of film heat transfer coefficients, consideration is limited to local coefficients α. Further, the above arguments demonstrate that the direct application of eqn. (19) is not appropriate in general and must be applied with caution even where the condensation of a pure vapour is considered.

3.3. Calculation of Heat Transfer Coefficients

In the application of modern design procedures the use of local coefficients is preferred to averaging coefficients over a finite tube length for the reasons suggested in Section (3.2). In the methods presented in this section, consideration is therefore limited to methods for prediction of local coefficients.

Condensate Layer

The original solution for laminar filmwise condensation on plane and curved surfaces is that of Nusselt.[14] For gravity controlled condensation on a vertical surface, flat or tubular, the coefficient α_l at a distance z from the top of the surface is given by

$$\alpha_l = \left\{ \frac{\lambda_l^3 \rho_l^2 g L}{3 \mu_l \dot{M}_l} \right\}^{1/3} \qquad \begin{matrix} \text{tubes} & L = d \\ \text{surfaces} & L = \text{width} \end{matrix} \tag{24}$$

while for condensation inside or outside a horizontal tube

$$\alpha_l = 0{\cdot}959 \left\{ \frac{\lambda_l^3 \rho_l^2 g L}{\mu_l \dot{M}_l} \right\}^{1/3} \qquad L = \text{length} \tag{25}$$

With the forced convective flows normally found in condensing equipment, the film thickness is affected by vapour shear, and the film itself usually becomes turbulent. Suitable methods for use in the design of tubular condensers are discussed below.

Condensation inside tubes: The two flow regimes of greatest practical interest with in-tube condensation are stratified and annular flow. These regimes are discussed in Section (2.2) and are depicted in Fig. 4.2. Stratified and annular flows correspond in horizontal tubes to gravity and shear-controlled film flows respectively. A flow pattern criterion has been proposed by Jaster and Kosky[15] in terms of a parameter F which reflects the ratio of shear to gravitational forces. For low-pressure condensation of steam their observations suggest:

$$\begin{matrix} F \geq 29 & \text{annular flow} \\ 5 < F < 29 & \text{transition flow} \\ F \leq 5 & \text{stratified-type flow} \end{matrix}$$

with,

$$F = \frac{\tau_w^{3/2}}{\rho_l^{1/2} g \mu_l s_F^+} \tag{26}$$

where τ_w is the wall shear stress, s_F^+ is a non-dimensional film thickness defined by

$$s_F^+ = \frac{(\tau_w \rho_l)^{1/2} s_F}{\mu_l} \tag{27}$$

and s_F is the annular film thickness. Recently Breber *et al.*[16] have found that the two-phase flow regime prediction models suggested by Taitel and

Dukler[17] apply to in-tube condensation. Breber *et al.* use the parameters j_g^*, defined by

$$j_g^* = \frac{\dot{M}_g}{S_T\sqrt{dg\rho_g(\rho_l - \rho_g)}} \tag{28}$$

and the Martinelli parameter X where,

$$X = \sqrt{\Delta P_l/\Delta P_g} \tag{29}$$

to produce a simplified flow map. The limits of the annular and stratified-type flows are given by

$j_g^* > 1{\cdot}5$ $\quad X < 1{\cdot}0$ $\quad$ annular flow

$j_g^* < 0{\cdot}5$ $\quad X > 1{\cdot}0$ $\quad$ stratified-type flow

For the transition zone between stratified and annular flow, Jaster and Kosky[15] suggest linear interpolation between the gravity-controlled and shear-controlled coefficients with respect to F, while Breber *et al.*[16] suggest interpolation with respect to j_g^*.

Models for the heat transfer coefficient in stratified flows assume that the Nusselt equation for curved surfaces applies to the region of the tube over which the film drains vertically or near-vertically. Heat transfer through the pool at the bottom of the tube can be neglected, and the models can be expressed as

$$\alpha_l = C(\alpha_l)_{\text{Nusselt}} \tag{30}$$

where C is a correction term. Palen *et al.*[18] recommend as a first approximation $C = 0{\cdot}79$, whilst Jaster and Kosky[15] relate the correction to the void fraction ε_g by

$$C = \varepsilon_g^{3/4} \tag{31}$$

where ε_g is calculated from the Zivi equation[19]

$$\varepsilon_g = 1\bigg/\left\{1 + \frac{1-x}{x}\left(\frac{\rho_g}{\rho_l}\right)^{2/3}\right\} \tag{32}$$

A simple approach to annular film condensation is the analogy relationship

$$\frac{\alpha_l}{\alpha_{lo}} = f\left(\frac{\Delta P_{TP}}{\Delta P_{lo}}\right) \tag{33}$$

where the subscript lo refers to the total mass flowrate flowing as liquid.

Using a homogeneous theory for the two-phase pressure drop ΔP_{TP}, Boyko and Kruzhilin[20] obtain

$$\alpha_l = 0{\cdot}021 \frac{\lambda_l}{d} \mathrm{Re}_{lo}^{0{\cdot}8} \mathrm{Pr}_l^{0{\cdot}43} \left\{1 + x\left(\frac{\rho_l}{\rho_g} - 1\right)\right\}^{1/2} \tag{34}$$

More accurate approaches involve a more direct evaluation of the film thickness and temperature drop. Kosky and Staub[21] integrate the laminar sublayer profile and the 1/7th power law velocity distribution from the wall to thickness s_F. Their results are:

$$\begin{aligned} s_F^+ &= (\mathrm{Re}_l/2)^{1/2} \qquad && \mathrm{Re}_l < 1000 \\ s_F^+ &= 0{\cdot}0504\,\mathrm{Re}_l^{7/8} \qquad && \mathrm{Re}_l > 1000 \end{aligned} \tag{35}$$

The Martinelli analogy is then applied to obtain the dimensionless film temperature drop T^+, given by

$$T^+ = (\rho_l \tau_w)^{1/2} c_{pl}/\alpha_l \tag{36}$$

as a function of the liquid Prandtl number and the dimensionless film thickness. Jaster and Kosky give

$$\begin{aligned} T^+ &= \mathrm{Pr}_l s_F^+ && s_F^+ \le 5 \\ T^+ &= 5[\mathrm{Pr}_l + \ln\{1 + \mathrm{Pr}_l(s_F^+/5 - 1)\}] && 5 < s_F^+ \le 30 \\ T^+ &= 5\left[\mathrm{Pr}_l + \ln(1 + 5\mathrm{Pr}_l) + \tfrac{1}{2}\ln\frac{s_F^+}{30}\right] && s_F^+ > 30 \end{aligned} \tag{37}$$

Other similar relations are given by Altman *et al.*,[22] Traviss *et al.*[23] and Bae *et al.*[24] A relationship earlier by Carpenter and Colburn[25] can be expressed as

$$T^+ = 23{\cdot}3\,\mathrm{Pr}_l \tag{38}$$

The wall shear stress τ_w is evaluated from the frictional component of the two-phase pressure gradient, calculated using standard methods for two-phase flow without phase change, and the gravitational term if necessary, that is

$$\tau_w = -\frac{d}{4}\left(\frac{\mathrm{d}P_{TP}}{\mathrm{d}z}\right)_{Fr} + (\rho_l - \rho_g) g s_F \sin\alpha \tag{39}$$

In the absence of reliable general correlations, the effect of waves on the condensate heat transfer coefficient is best ignored, and treated as a safety factor.

Condensation on tube bundles: Considerably less is known about forced convective condensation outside tube bundles. The approach usually adopted is to correct the basic Nusselt prediction separately for the effects of inundation and vapour shear. The inundation effect is often expressed by an equation of the form for α_l of the Nth row, α_{lN}

$$\alpha_{lN} = \left\{\frac{W + w}{W}\right\}^{-\gamma} (\alpha_{lT})_{\text{Nusselt}} \tag{40}$$

where $(\alpha_{lT})_{\text{Nusselt}}$ is the Nusselt coefficient for the top row, W is the condensate inundation rate on the Nth row and w is the condensation rate on the Nth row. Reported values of γ range from 0·07 to 0·225. A value of 0·16 was found by Wilson,[26] using a two-dimensional analysis, based on the data of Grant and Osment[27] from a realistically sized test condenser.

The effect of vapour shear has been considered at various levels of sophistication. The simplest equation is that of Shekriladze and Gomelauri[28] who consider only the shear forces due to momentum transferred by the condensing vapour to the condensate film. Their equation for the top row is

$$\alpha_{lT} = \left\{0{\cdot}405\left(\frac{\lambda_l^2 \rho_l \mu_g}{\mu_l d}\right) + \left[0{\cdot}164\left(\frac{\lambda_l^2 \rho_l \mu_g}{\mu_l d}\right)^2 + (\alpha_{lT})_{\text{Nusselt}}\right]^{1/2}\right\}^{1/2} \tag{41}$$

Fujii *et al.*[29] extend this solution to include a frictional shear stress term. Recently, more advanced solutions, by Nicol and Wallace,[30] Nicol *et al.*,[31] Nobbs and Mayhew[32] and Fujii *et al.*[33] have been proposed. These solutions involve numerical integrations to evaluate the circumferential film thickness distribution around the tube, and are not suitable for design purposes.

The combined, interactive, effects of vapour shear and inundation are not currently represented by any published design correlation.

Condensation of Vapours of Immiscible Liquids

In certain process applications the condensate forms two immiscible liquid phases, commonly encountered with condensation of steam-hydrocarbon systems. The condensate film resistance then depends on the distribution of the phases over the cooling surface. The two most common flow patterns observed are channelling flow and a standing drop pattern though no method is currently available for determining the flow pattern which will be realised under given process conditions. A number of experimental studies have been reported, in all of which the condensate was in laminar flow and gravity controlled. These are summarised by Bernhardt *et al.*,[34] who proposed

$$\alpha_1 = \alpha_{11} V_1 + \alpha_{12} V_2 \tag{42}$$

where α_{11} and V_1 are the Nusselt film coefficients and volume fractions of phase 1 in the condensate. Bernhardt reports that this equation predicted the majority of data from various sources to within $\pm 20\,\%$. Another widely used model is that of Akers and Turner.[35]

More recently Polley and Calus[36] have analysed the liquid film structure in more detail than previous authors, but methods of predicting liquid film coefficients remain uncertain.

Methods for dealing with inundation and vapour shear in immiscible liquid condensation are extremely uncertain.

Vapour Film Resistance

In condensation from a mixture of vapours with or without a non-condensing gas, simultaneous heat and mass transfer occurs across the temperature and concentration gradients between the bulk gas phase and the condensate surface. The only practicable design methods to evaluate the corresponding heat and mass transfer coefficients are based on film theory.

The Nusselt number for heat transfer from a gas to a dry wall in turbulent flow inside or outside a tube bundle is well represented by correlations of the form

$$\mathrm{Nu} = \mathrm{a}_1 \, \mathrm{Re}_{\mathrm{g}}^{\mathrm{a}_2} \, \mathrm{Pr}_{\mathrm{f}}^{\mathrm{a}_3} \tag{43}$$

where a_1, a_2 and a_3 are numerical constants depending on the geometry. This can be expressed in terms of the *j*-factors of Chilton and Colburn[37] with numerical constants b_1, b_2 and b_3 as

$$j_{\mathrm{H}} = \frac{\alpha_{\mathrm{g}} S}{c_{\mathrm{pf}} \dot{M}_{\mathrm{g}}} \mathrm{Pr}_{\mathrm{f}}^{\mathrm{b}_3} = \mathrm{b}_1 \, \mathrm{Re}_{\mathrm{g}}^{\mathrm{b}_2} \tag{44}$$

The analogy between heat and mass transfer as suggested by the same authors gives

$$j_{\mathrm{D}} = \frac{\beta_{\mathrm{g}ij} S}{\dot{N}_{\mathrm{g}}} (\mathrm{Sc}_{ij})_{\mathrm{f}}^{\mathrm{b}_3} = \mathrm{b}_1 \, \mathrm{Re}_{\mathrm{g}}^{\mathrm{b}_2} \tag{45}$$

These relationships for j_{H} and j_{D} for transfer to a dry wall, form the basis of design based on film theory. When condensation transfer occurs to a moving surface, which can be rippled or wavy, and the mass transfer rate is high it is almost certain that the pattern of the turbulence close to the surface is affected. The effect of surface waves has been thought to increase the coefficients, but no direct experimental evidence is yet available. Price and Bell[38] have suggested a correction to the dry-gas heat and mass transfer coefficients.

The effect of high mass transfer rates towards the surface is estimated using film theory. A general presentation, applicable to momentum, heat and mass transfer is given by Bird, Stewart and Lightfoot.[39] Their results correspond to eqns. (5) and (7).

Berman[40] presents correlations of experimental data which offer an empirical alternative to film theory. Over a wide concentration range Berman's correction term for mass transfer differs from film theory by at most 12%.

In multicomponent condensation there is often considerable variation in the physical and transport properties across the gas film. This is often caused more by concentration than by temperature differences, and the variations are generally significant when the gas-phase resistance is controlling. In the use of such correlations in this work, it has been the policy to evaluate the mixture properties involved in the Prandtl and Schmidt numbers at the arithmetic mean of the bulk gas phase and surface temperature and concentration (e.g. $\mathrm{Pr_f}$), and those involved in the Reynolds number at bulk gas-phase conditions ($\mathrm{Re_g}$). However where specific recommendations are made for a given correlation these should be followed.

Interfacial Resistance

On the basis of kinetic theory, a resistance to heat transfer might be expected to occur at the vapour–liquid interface. However, this resistance is usually so small compared with other resistances that it can be neglected in design, and the assumption that the vapour and liquid at the interface are at the same temperature is therefore valid. The interfacial resistance can be significant, however, when condensing a pure vapour at very low pressure (say less than 0·03 bar). The interfacial coefficient α_I, can be evaluated from an equation of Berman[41]:

$$\alpha_\mathrm{I} = \frac{2\sigma_\mathrm{c}}{2 - \sigma_\mathrm{c}} \left\{ \frac{\tilde{M}}{2\pi RT} \right\}^{1/2} \frac{\Delta h_\mathrm{v}^2 P \tilde{M}}{RT^2} \tag{46}$$

where σ_c is the condensation coefficient. Wide variations in experimental measurements of σ_c have been reported, probably reflecting difficulties of measurement.

Coolant Film Resistance
Correlations of the form of eqn. (43) are available for a wide range of geometries and are used to calculate the heat transfer coefficient of the hypothetical coolant film α_c^*. The resistance of the wall, and of dirt and fouling layers that might be expected from operating experience with the fluids involved, are lumped to form an overall heat transfer coefficient α_c, from the vapour side of the wall to the coolant.

$$\frac{1}{\alpha_c} = \underset{\text{(dirt)}}{\frac{1}{\alpha_d}} + \underset{\text{(wall)}}{\frac{1}{\alpha_w}} + \underset{\text{(coolant)}}{\frac{1}{\alpha_c^*}} \tag{47}$$

3.4. Transport and Thermodynamic Property Data in Design
In this section the fluid properties necessary for condenser design are listed and in the case of hydrocarbons the sources of property data used in the design examples are quoted. A very comprehensive set of properties are required and the assembly of a self-consistent and reliable physical property data bank is essential for condenser design. Physical properties are in general temperature, pressure and composition dependent and methods of calculation of the effects of process conditions must be available. In the more sophisticated computer design methods of Section (4), the repeated evaluation of physical properties during the design is a major consumer of computer time. Unfortunately, there is no guidance in the literature as to the sensitivity of the various design variables to the physical properties, and the individual designer must rely on his own intuition. The following physical and transport properties are always required for coolant, condensate and vapour mixtures, irrespective of the design method:

(i) Density ρ (kg/m^3).
(ii) Specific heat capacity c_p (J/kg K).
(iii) Viscosity μ (Ns/m^2).
(iv) Thermal conductivity λ (W/m K).
(v) Enthalpy h (J/kg).
(vi) Molecular weight $\tilde{M}$ (kg/kmol).

In addition vapour–liquid equilibrium calculations have to be carried out so that equilibrium ratios, or 'K' values must be available, where,

$$K_i = \tilde{y}_i / \tilde{x}_i$$

In the mechanistic design methods, where mass transfer is separately described, binary diffusivities must be calculated for all constituent binary pairs of the mixture. In addition, the latent heat of vaporisation is required and this is dependent on the partial molar enthalpy differences between vapour and liquid $\Delta\tilde{h}_{vi}$.

In the examples of Section (4) hydrocarbon mixtures are considered. The above data may be assembled from the following sources:

(i) Vapour–liquid equilibrium is described by the Chao–Seader method,[42] which uses the theory of regular solutions to describe liquid behaviour and the Redlich--Kwong equation of state in the vapour. Additional information is to be found in the work of Cavett[43] and Grayson and Streed[44] while Shelton[45] describes the development of computational procedures.

(ii) Ideal gas enthalpies may be obtained from the extensive data in the API Technical Data Book[46] and corrections are applied for pressure within the framework of regular solution theory and the Redlich–Kwong equation to give enthalpies of vapour and liquid mixtures and partial molar enthalpies of the components of these mixtures. (See for example Edmister.[47]) Estimates of specific heat capacities follow directly from the calculated enthalpies. Liquid and vapour densities may be calculated on the same basis.

(iii) Transport properties may be calculated by the methods summarised by Reid and Sherwood.[48] In the calculations of Section (4) the vapour–gas mixture properties were estimated as follows; thermal conductivities by the methods of Misic and Thodos (see ref. 48, p. 466), binary diffusivities (see ref. 48, p. 523) and viscosities by the methods of Stiel and Thodos (see ref. 48, p. 404). Liquid viscosities and thermal conductivities were obtained from experimental data fitted by appropriate equations (see ref. 48, pp. 431 and 505).

4. THE DESIGN OF CONDENSERS OF SIMPLE GEOMETRY

Design methods may be broadly classified into two categories. These are equilibrium methods, typified by the procedures of Kern[49] or of Silver[50] (as modified by Bell and Ghaly),[51] and the differential methods which have developed from the original work of Colburn and Drew.[7]

In the equilibrium methods, no attempt is made to describe the process of vapour-side mass transfer. The bulk vapour and liquid condensate at any plane normal to the flow are assumed to be at equilibrium at a temperature intermediate to the dew and bubble points of the feed. The method is particularly well suited to the situation in which vapour and condensate do not become separated, since in this case the overall local composition is the same as the vapour feed composition.

In the second class of methods, the equations of Section (3.1), which allow the calculation of local heat and mass transfer rates, are combined with differential mass and energy balances, which describe the downstream development of the independent vapour and coolant temperatures and vapour composition through the condenser. Design is effected by integration of these differential equations.

Equilibrium methods have the advantages of simplicity and speed and are suitable for hand calculation. No binary diffusivity data are required since mass transfer processes are not directly involved. A price must be paid for this simplicity. The assumption of local equilibrium prevents the prediction of effects such as fogging, which are the result of departures from equilibrium, and in condensation from a non-condensable gas the outlet composition may well be very different from the equilibrium composition. The application of the equilibrium methods becomes more involved where vapour–condensate separation occurs.[52]

Equilibrium methods and film methods of design are considered in Sections (4.1) and (4.2) respectively. Each method is followed by an example to assist in understanding and to act as a sample calculation. Vapour, coolant and condensate are assumed to be in cocurrent downwards flow to highlight the design method itself. Design in more complicated geometries is discussed in Section (5).

4.1. Equilibrium Design Methods

Let us suppose that in a condenser the bulk vapour and liquid condensate at a plane normal to the direction of vapour flow are considered to be at equilibrium at the bulk vapour temperature. Then the dew point T_{dew} of the vapour will correspond with the onset of condensation and the bubble point T_{bub} will correspond with total condensation. At temperatures above the dew point, de-superheating of the vapour occurs and at temperatures below the bubble point liquid subcooling occurs.

At any temperature intermediate to the dew and bubble points the composition of the equilibrium liquid $\tilde{x}_i$ and vapour $\tilde{y}_i$ may be determined by vapour–liquid equilibrium considerations. Thus the vapour fraction $\tilde{\theta}$ must satisfy the following, if the vapour feed composition is $\tilde{Z}_i$ (assuming no separation of condensate and vapour) then

$$\tilde{Z}_i = \tilde{\theta}\tilde{y}_i + (1 - \tilde{\theta})\tilde{x}_i \qquad i = 1, n \tag{48}$$

where,

$$\tilde{y}_i = K_i \tilde{x}_i \qquad i = 1, n \tag{49}$$

solving for x_i,

$$\tilde{x}_i = \tilde{Z}_i / \{1 + \tilde{\theta}(K_i - 1)\} \qquad i = 1, n \tag{50}$$

A temperature T is sought such that

$$\sum_{i=1}^{n} \tilde{y}_i - \sum_{i=1}^{n} \tilde{x}_i = 0 \tag{51}$$

At this temperature, consideration of eqn. (48) summed over n components leads to the result that

$$\sum^{n} \tilde{y}_i = 1{\cdot}0 \qquad \text{and} \qquad \sum^{n} \tilde{x}_i = 1{\cdot}0$$

and $\tilde{y}_i$ and $\tilde{x}_i$ at this stage are the compositions of the equilibrium vapour and liquid. The search for the appropriate value of $\tilde{\theta}$ is straightforward because it can be shown that $\sum \tilde{y}_i - \sum \tilde{x}_i$ is a decreasing function of $\tilde{\theta}$ at constant temperature. The above is of course the well known method of calculation of vapour fraction, but it will have special significance in the methods proposed later in this section and has been included for that reason.

The condensation, or cooling curve, (see Fig. 4.4) is now prepared. This is a plot of vapour temperature T_g, against cumulative heat load $\dot{Q}$. The above calculations must be carried out at a number of selected temperatures

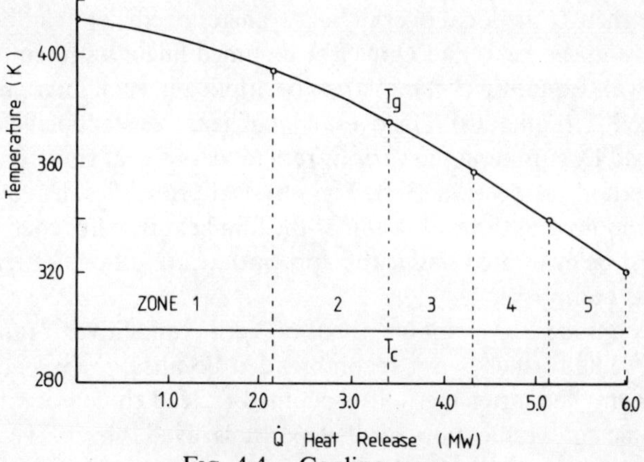

FIG. 4.4. Cooling curve.

between T_{dew} and T_{bub} and the molar specific enthalpies of vapour and liquid $\tilde{h}_g$ and $\tilde{h}_l$ respectively, determined at these temperatures. The stream average molar specific enthalpy $\tilde{h}$ is calculated and hence knowing the molar vapour feed rate $(\dot{N}_g)_{in}$ and the molar specific enthalpy at the vapour inlet $\tilde{h}_{in}$, $\dot{Q}$ may be evaluated.

$$\tilde{h} = \tilde{\theta}\tilde{h}_g + (1 - \tilde{\theta})\tilde{h}_l \tag{52}$$

$$\dot{Q} = (\dot{N}_g)_{in}(\tilde{h}_{in} - \tilde{h}) \tag{53}$$

The variation of cooling fluid temperature T_c with $\dot{Q}$ is included with the cooling curve to show the variation of temperature driving force $\Delta T = (T_g - T_c)$ with $\dot{Q}$.

The temperature T_c is calculated from the coolant flowrate $\dot{M}_c$, and specific heat c_{pc}, taking account of whether flow is cocurrent or countercurrent. The discussion of how to handle more complicated flow arrangements such as multiple fluid passes is deferred to Section (5.1).

$$\begin{aligned} \dot{Q} &= \dot{M}_c c_{pc}\{(T_c)_{out} - T_c\} && \text{countercurrent} \\ &= \dot{M}_c c_{pc}\{T_c - (T_c)_{in}\} && \text{cocurrent} \end{aligned} \tag{54}$$

The local molar flowrates of vapour and condensate $\dot{N}_g$ and $\dot{N}_l$, are readily calculated from $\tilde{\theta}$ (as are mass flowrates $\dot{M}_g$):

$$\begin{aligned} \dot{N}_g &= (\dot{N}_g)_{in}\tilde{\theta} = \dot{M}_g/\tilde{M}_g = (\dot{M}_g)_{in}x/\tilde{M}_g \\ \dot{N}_l &= (\dot{N}_g)_{in}(1 - \tilde{\theta}) = \dot{M}_l/\tilde{M}_l = (\dot{M}_g)_{in}(1 - x)/\tilde{M}_l \end{aligned} \tag{55}$$

where $x = \tilde{\theta}\tilde{M}_g/M_{in}$ is the vapour mass fraction or quality.

If the number of temperature intervals is sufficient (five are shown in Fig. 4.4), then U the local overall heat transfer coefficient can be assumed constant, and the heat load $\dot{Q}$ may be assumed linear with temperature in that interval. Equation (14) may then be applied in each interval with ΔT_m given by ΔT_{1n}, (eqn. (20)). The overall local heat transfer coefficient U may be obtained by summing the various resistances to heat transfer, predicted by the methods of Section (3.3). The physical properties, listed in Section (3.4) are required in the evaluation of the film heat transfer coefficients and these must be evaluated under the appropriate conditions of temperature and phase composition.

Two techniques are available, those of Kern[49] and Silver[50] (and Bell and Ghaly).[51] The former is not recommended because it can lead to severe underdesign. The latter method is claimed to lead to conservative design but insufficient verification of this point is available in the literature. Certainly however the method is very widely applied in current engineering design practice.

Kern Method (1950)
It is assumed that the full heat load $\Delta\dot{Q}$ in each interval of area ΔA must be conducted across the condensate film and the composite resistance of wall, dirt layers and coolant film. With eqns. (1), (3), (20), (24) and (47) it follows that

$$\Delta A = \Delta\dot{Q}/(\Delta T_{1n} U) \tag{56}$$

where

$$U = \alpha_c \alpha_l/(\alpha_c + \alpha_l) \tag{57}$$

where liquid condensate properties are evaluated at the mean liquid temperature given by

$$T_l = T_c + \tfrac{1}{2}\left\{\frac{U}{\alpha_c} + 1\right\}(T_g - T_c) \tag{58}$$

Silver Method (1947)
Once again it is assumed that the total heat load $\Delta\dot{Q}$ must be conducted across the various resistances between the vapour–liquid interface and the coolant. However, in addition a sensible heat load $\Delta\dot{Q}_s$ is distinguished and it is assumed that this must pass by conduction across the vapour film, the heat transfer coefficient of which α_g is predicted by the methods of Section (4.3c), with physical properties evaluated at the mean film temperature T_f.

Defining,

$$\Delta\dot{Q}_s = -\dot{N}_g \tilde{c}_{pg} \Delta T_g \tag{59}$$

and,

$$Z = \frac{\Delta\dot{Q}_s}{\Delta\dot{Q}} = \dot{N}_g \tilde{c}_{pg}\left\{-\frac{\Delta T_g}{\Delta\dot{Q}}\right\} \tag{60}$$

The quantity $(-\Delta T_g/\Delta\dot{Q})$, intrinsically a positive quantity, is the negative of the slope of the cooling curve in the interval under consideration. It is easily verified that the local overall heat transfer coefficient U and the temperatures T_l and T_f, are given by

$$U = 1\Bigg/\left\{\frac{Z}{\alpha_g} + \frac{1}{\alpha_c} + \frac{1}{\alpha_l}\right\} \tag{61}$$

$$T_l = T_c + \frac{U}{2\alpha_c}\left\{2 + \frac{\alpha_c}{\alpha_l}\right\}(T_g - T_c) \tag{62}$$

$$T_f = T_c + \tfrac{1}{2}\left\{\frac{U(\alpha_c + \alpha_l)}{\alpha_l \alpha_c} + 1\right\}(T_g - T_c) \tag{63}$$

Equations (58), (62) and (63) are the mean film temperatures of condensate and vapour and a logical choice of temperature for the evaluation of physical properties. However they cannot be determined *a priori* since they depend on the various heat transfer coefficients, which in turn depend on the mean film properties and therefore temperature. Hence an iterative procedure must be adopted if they are to be used, ruling out the possibility of hand calculation. In the absence of specific recommendations in the literature, and experimental data to back up the use of these equations, and bearing in mind the uncertainty of the model itself, it is hard to justify the evaluation of physical properties at any temperature other than T_g. Further, in actual design the error involved in evaluating physical properties at the bulk vapour temperature may be small. (In Example 1, which follows, the calculated numbers of tubes required for the given duty are within 5%.)

Example 1. A saturated mixture of hydrocarbons (43 386 kg/h) having the composition given below are to be condensed in downwards flow through the tubes (details below) of a shell-and-tube heat exchanger.

Component	*Composition* $\tilde{y}_{ig}$	*Tube data*
Propane (C_3H_8)	0·15	Inside diameter d (0·025 4 m)
n-Butane (C_4H_{10})	0·25	Tube length (2·438 m)
n-Hexane (C_6H_{14})	0·05	Area A_T (0·194 5 m^2)
n-Heptane (C_7H_{16})	0·30	Coolant temperature (300 K)
n-Octane (C_8H_{18})	0·25	Heat transfer coefficient α_c (1 700 W/m^2 K)

The coolant temperature is to be assumed constant at 300 K and the heat transfer coefficient between wall and bulk coolant is to be assumed constant at 1700 W/m^2 K. Calculate (i) the number of tubes required using the Kern method, and (ii) the extra tube length required to allow for sensible heat transfer using the Silver method with that number of tubes. (Answer (i) 1324 tubes, (ii) tube length doubled.)

Solution

Step 1. Calculate the dew and bubble points T_{dew} and T_{bub} of the mixture by finding a set of $\tilde{y}_i$ and $\tilde{x}_i$ which satisfy eqns. (48), (49) and (51) with $\tilde{\theta} = 1{\cdot}0$ and 0 respectively. (Table (4.2).)

TABLE 4.2

PHYSICAL PROPERTIES IN EXAMPLE 1

Property	*Position 1* (*dew point*) T_{dew}	*Position 2*	*Position 3*	*Position 4*	*Position 5*	*Position 6* (*bubble point*) T_{bub}
Temperature (K)	413·1	394·8	376·5	358·2	339·9	321·6
Liquid properties						
Density (kg/m^3)	456·4	482·2	511·2	541·4	569·9	594·6
Specific heat capacity (J/kg K)	2 385·0	2 244·0	2 115·0	2 009·0	1 900·0	1 803·0
Viscosity ($\times 10^3$) (Ns/m^2)	0·173 5	0·187 9	0·204 8	0·213 6	0·227 4	0·238 7
Thermal conductivity (W/m K)	0·111 4	0·114 0	0·116 6	0·119 2	0·121 8	0·124 4
Vapour properties						
Density (kg/m^3)	9·075	8·182	7·555	7·189	7·011	6·928
Specific heat capacity (J/kg K)	2 201·0	2 140·0	2 079·0	2 016·0	1 950·0	1 880·0
Viscosity ($\times 10^5$) (Ns/m^2)	0·897	0·907	0·908	0·897	0·878	0·853
Thermal conductivity ($\times 10^1$) (W/m K)	0·155 1	0·162 9	0·165 2	0·150 4	0·143 9	0·136 3
Vapour fraction $\tilde{\theta}$	1·000 0	0·619 1	0·425 9	0·292 3	0·161 2	0·000 0
Quality $x = \tilde{\theta}\tilde{M}_g/\tilde{M}_{in}$	1·000 0	0·538 4	0·328 2	0·204 2	0·104 2	0·000 0
Heat released $\dot{Q}$ (kW)	0·0	2 154·0	3 403·0	4 325·0	5 153·0	5 995·0

Step 2. Divide the temperature range ($T_{dew} - T_{bub}$) into five equal intervals and at each intermediate temperature find the $\tilde{\theta}$ which satisfies eqns. (48) to (51) (Table (4.2)). (The vapour mass fraction, or quality x, is also calculated.)
Step 3. For the six temperatures defined evaluate the density, specific heat capacity, enthalpy, viscosity and thermal conductivities of the equilibrium vapour and liquid at temperature T_g using the methods of Section (3.4) (Table (4.2)).
Step 4. Calculate the cumulative heat load, or heat released $\dot{Q}$, for each interval by eqns. (52) and (53) and plot the condensation curve T_g against $\dot{Q}$. (See Fig. 4.4.)
Step 5. Calculate the coolant temperature from eqn. (54) and plot on Fig. 4.4. Calculate the mass flowrates $\dot{M}_g$ and $\dot{M}_l$, and molar flowrates $\dot{N}_g$ and $\dot{N}_l$, of vapour and condensate, from eqn. (55). (In this example T_c is taken as constant.)
Step 6. Calculate arithmetic average values of gas and coolant temperatures, and gas and condensate physical properties and flowrates in each interval. (All gas and condensate physical properties are evaluated at T_g.)
Step 7. Guess the number of tubes required.
Step 8. On the basis of a single tube calculate the flowrates $\dot{M}_g$, and heat loads $\Delta\dot{Q}$ and $\Delta\dot{Q}_s$ in each interval. The heat load $\Delta\dot{Q}$ is specified if $\dot{Q}$ is known (Table (4.2)) and $\Delta\dot{Q}_s$ is given by eqn. (59). Calculate Z from eqn. (60). (Table (4.3).)
Step 9. (*Optional*) If desired, calculate appropriate physical properties at the film average and condensate average temperatures T_f and T_l, given by eqns. (62) and (63), once values of α_g, α_c and α_l are available.
Step 10. In each interval calculate the heat transfer coefficients of vapour film α_g, condensate α_l, and the overall liquid side α_c, by eqns. (44), (24) and (47) respectively. In eqn. (44), $b_1 = 0{\cdot}023$, $b_2 = -0{\cdot}17$, $b_3 = 2/3$. (Table (4.3), in this example α_c is specified.)
Step 11. Calculate the overall heat transfer coefficient U in each interval by eqn. (61), and the log mean temperature difference ΔT_m, by eqn. (20). If Step 9 is included repeat Steps 9–11 to convergence. (Kern method—evaluate U with $Z = 0$.)
Step 12. Calculate the incremental area ΔA required for condensation in each interval from eqn. (56) and sum to give A_T. Calculate the total exchanger area and the number of tubes required to give this area.
Step 13. Repeat Steps 8–12 until the number of tubes is converged. (Table (4.3) gives the converged solution based on a single tube.)

TABLE 4.3
DESIGN CALCULATIONS IN EXAMPLE 1

Basis (1 tube)		*Interval 1*	*Interval 2*	*Interval 3*	*Interval 4*	*Interval 5*
Flowrates						
Gas flowrate $\dot{M}_g$ (kg/s)		0·007 004	0·003 945	0·002 424	0·001 404	0·000 474
Condensate flowrate $\dot{M}_l$ (kg/s)		0·002 101	0·005 160	0·006 681	0·007 701	0·008 361
Calculation of α_g						
Reynolds number Re_g		38 922	21 786	13 458	7 932	2 749
Chilton–Colburn J_H factor		0·003 81	0·004 21	0·004 57	0·005 00	0·005 99
Prandtl number Pr		1·232	1·168	1·171	1·196	1·182
Gas film heat transfer coefficient α_g (W/m^2 K)		99·60	62·35	40·27	24·38	9·59
Liquid film heat transfer coefficient α_l (W/m^2 K)		600·7	460·1	439·2	436·1	434·4
Energy transfer						
Total $\Delta\dot{Q}$ (W)		1 627·5	943·4	697·0	625·3	635·8
Sensible $\Delta\dot{Q}_s$ (W)		278·0	152·2	90·8	50·9	16·61
Z		0·170 8	0·161 3	0·130 3	0·081 4	0·026 1
Overall heat transfer coefficient U (W/m^2 K)	(Kern)	443·6	361·8	348·7	346·8	345·7
	(Silver)	252·0	187·0	164·0	160·7	178·2
Log mean temperature difference ΔT_{ln} (K)		103·6	85·3	66·9	48·5	29·8
Area required ΔA (m^2)	(Kern)	0·035 4	0·030 6	0·029 9	0·037 2	0·061 7
	(Silver)	0·062 3	0·059 2	0·063 5	0·080 2	0·119 6

From Table (4.3), it is seen that the Kern method (with $Z = 0$) predicts that 0·009 105 kg/s of the feed can be condensed in a tube of area 0·1948 m^2. Since this agrees with the stated tube area of 0·1945 m^2, it follows that 1324 tubes are required to condense the given flow. With this number of tubes the Silver method predicts the area required to be 0·3848 m^2—almost twice that of the Kern method.

4.2. Film Models of Design

The equations for local conditions at a plane normal to the vapour flow have been presented in full in Section (3.1). In general, vapour composition $\tilde{y}_{ig}$, temperature T_g and coolant temperature T_c, are known and the molar rates of transfer $\dot{n}_i$, and heat flux $\dot{q}_I$, must be determined. These quantities may only be calculated by the equations of Section (3), once the interfacial state is known. The interfacial temperature T_I, must lie between T_g and T_c, and χ (which must lie between 0 and 1) is defined as the dimensionless interfacial temperature,

$$\chi = \frac{(T_I - T_c)}{(T_g - T_c)} \tag{64}$$

Defining α_{ol} as the combined heat transfer coefficient of condensate and wall (including fouling and coolant films), eqns. (1), (3), (5) and (8) may be combined to give,

$$\dot{n}_T = \frac{\alpha_{ol}\chi}{\left\{\dfrac{\Delta\tilde{h}_v}{(T_g - T_c)} + \dfrac{\tilde{c}_p(1-\chi)}{[1 - \exp(-\varepsilon)]}\right\}} \tag{65}$$

where,

$$\Delta\tilde{h}_v = \sum_{i=1}^{n} \frac{\dot{n}_i}{\dot{n}_T} \Delta\tilde{h}_{vi}$$
$$\tilde{c}_p = \sum_{i=1}^{n} \frac{\dot{n}_i}{\dot{n}_T} \tilde{c}_{pi} \tag{66}$$

and

$$\alpha_{ol} = 1{\cdot}0 \bigg/ \left\{\frac{1}{\alpha_I} + \frac{1}{\alpha_c}\right\} \tag{67}$$

With assumed physical properties $\Delta\tilde{h}_v$ and $\tilde{c}_p$, eqn. (65) may be readily solved when χ is known to give the total condensation flux $\dot{n}_T$. An iterative procedure is involved because ε is a function of $\dot{n}_T$ but eqn. (65) is readily solved by repeated substitution using $\varepsilon = 1{\cdot}0$ as the initial guess. It is clear that $\dot{n}_T$ is constrained to have a maximum value when $\chi = 1$ and this lends stability in subsequent calculations.

At this stage the method of calculation will diverge depending on whether the condensate is assumed to be totally mixed or unmixed. Whichever assumption is made the initial conditions at the vapour inlet correspond to the case of liquid unmixed. It is the unmixed case which is the more difficult. The two are considered in turn.

4.2.1. *Condensate Liquid of Known Composition Present and Completely Mixed*

The condensate at any position through a condenser in a non-separating flow will be made up of all prior condensed species and will be of known composition $\tilde{x}_{il}$ which will also be the surface composition. The interfacial temperature T_I (and hence χ by eqn. (64)) and equilibrium vapour composition $\tilde{y}_{iI}$, are determined through a bubble point calculation which of course satisfies eqn. (4). The mean vapour film temperature T_f, and mean condensate liquid temperature T_l, are respectively

$$T_f = T_c + (1 + \chi)(T_g - T_c)/2{\cdot}0 \tag{68}$$

$$T_l = T_c + \chi\left\{1 + \frac{2\alpha_l}{\alpha_c}\right\}(T_g - T_c)\Big/2\left\{1 + \frac{\alpha_l}{\alpha_c}\right\} \tag{69}$$

Liquid condensate physical properties must be evaluated at T_l and vapour film properties at T_f, respectively. Clearly, T_f is defined, but T_l is dependent on the value of α_l which is not known *a priori*.

The following computational algorithm has proved successful in practice. Points of divergence between methods are indicated by KS (Krishna–Standart), LT (linearised theory) and ED (effective diffusivity).

Step 1. Evaluate gas film properties at T_g, $\tilde{y}_{ig}$ (Section (3.4)).
From the known $\tilde{x}_{iI} = \tilde{x}_{il}$, calculate T_I and $\tilde{y}_{iI}$ with $\tilde{\theta} = 0$. (Bubble point calculation.) (Equations (48, 49, 51).)
Step 2. Calculate χ (eqn. (64)).
Step 3. Calculate T_f (eqn. (68)).
Estimate T_l as T_I.
Evaluate vapour film properties at T_f, $(\tilde{y}_{iI} + \tilde{y}_{ig})/2{\cdot}0$.
Step 4. Evaluate liquid film properties at T_l, $\tilde{x}_{il}$ (Section (3.4)).

Step 5. Evaluate α_l, α_g, α_c (eqns. (24, 44 and 47)).
Calculate T_l (eqn. (69)).
Repeat Steps 4 and 5 to convergence.
Step 6. Estimate $\Delta\tilde{h}_v$ and $\tilde{c}_p$ as interfacial liquid composition averages $\tilde{x}_{il}$.
Put $\varepsilon = 1$.
Step 7. Calculate $\dot{n}_T$ (eqns. (65, 67)).
Step 8. Calculate ε (eqn. (5)).
Repeat steps 7 and 8 to convergence.
Step 9. Calculate the $\beta_{gij}(i,j = 1,n)$ (eqn. (45)).
Step 10. ED: Calculate ϕ_i (eqn. (7)).
KS: First time through use LT approach.
Calculate $[B]$ and $[\Phi]$ (Table (4.1)).
LT: Calculate $[B]$ and $[\Phi]$ (Table (4.1)).
Step 11. ED: Calculate j_{ig} (eqn. (10)).
KS and LT: Calculate j_{ig} (eqn. (9)).
Step 12. Calculate $\dot{n}_i$ (eqn. (6)).
Step 13. Evaluate $\Delta\tilde{h}_v$ and $\tilde{c}_p$ (eqn. (66)).
Repeat steps 7–13 to convergence.
Step 14. Evaluate $\dot{q}_l$ (eqns. (1, 3)).

It is worth noting that the iteration loop on ε is necessary only because eqn. (65) cannot be written explicitly for $\dot{n}_T$. The iteration loop on T_l becomes unnecessary if constant physical properties are assumed. The loop on $\Delta\tilde{h}_v$, $\tilde{c}_p$ (and $\dot{n}_i$) assumes special significance in the Krishna–Standart approach where matrix $[\Phi]$ is dependent on the individual $\dot{n}_i$, but is unnecessary with the linearised theory and effective diffusivity methods, if constant physical properties are assumed.

4.2.2. Initial Condensation or Liquid Condensate Present but Unmixed

This is a far more difficult calculation because the interfacial conditions are not defined. Equations (4), (6) and (9) or (10) may be solved with the unmixed liquid condition of eqn. (2) to give expressions for $(\tilde{x}_l)$ in terms of $(\tilde{y}_g)$. For the effective diffusivity model,

$$\tilde{x}_{il} = \frac{(1 + \beta^{\cdot}_{gin}/\dot{n}_T)\tilde{y}_{ig}}{(1 + K_i\beta^{\cdot}_{gin}/\dot{n}_T)} \qquad i = 1, n-1 \tag{70}$$

while for the matrix model,

$$(\tilde{x}_l) = \left\{\ulcorner I \lrcorner + \frac{1}{\dot{n}_T}[B^{\cdot}]\ulcorner K \lrcorner\right\}^{-1}\left\{\ulcorner I \lrcorner + \frac{1}{\dot{n}_T}[B^{\cdot}]\right\}(\tilde{y}_g) \qquad i = 1, n-1 \tag{71}$$

In both cases the nth composition is determined by difference

$$\tilde{x}_{nI} = 1 \cdot 0 - \sum_{i=1}^{n-1} \tilde{x}_{iI} \tag{72}$$

The corresponding vapour composition is of course given by eqn. (4).

If eqns. (50) and (70) or (71) are compared a striking similarity is apparent. Moreover $\tilde{\theta}$ and χ are both between 0 and 1. Clearly eqn. (51) must be satisfied by $\tilde{y}_{iI}$ and $\tilde{x}_{iI}$. This similarity is exploited in proposing an algorithm of similar structure to the calculation of vapour fraction.

Two nested iteration loops are constructed. The outer loop is on dimensionless temperature χ and the inner loop is essentially on composition but in the case of the Krishna–Standart method involves the iterative sequence necessary to evaluate the high flux corrections which are dependent on the molar fluxes $\dot{n}_i$ not known *a priori*. The criterion of convergence of the inner loop is that $\sum_{i=1}^{n} \tilde{y}_i$ is constant, and of overall convergence is that $\sum_{i=1}^{n} \tilde{y}_i = 1 \cdot 0$. The value of $\sum_{i=1}^{n} \tilde{y}_i$ is an increasing function of χ because the equilibrium K values of eqn. (4) increase with temperature. Practice has shown that the algorithm may be converged by selecting $\chi = 1 \cdot 0$ as the initial guess and decreasing χ in steps of 0·2 until the value of $\sum \tilde{y}_i - 1 \cdot 0$ changes sign. Thereafter, successive χ values are estimated by linear interpolation.

The convergence of the inner loop need not be to high precision if $\sum \tilde{y}_{iI}$ is very different from unity. An acceptable precision is $(\sum \tilde{y}_{iI} - 1 \cdot 0)/10 \cdot 0$ in successive estimates of $\sum \tilde{y}_{iI}$ until the required precision, here taken as 10^{-5}, is achieved. If the value of χ is chosen to be too low $\dot{n}_T$ will be small and it is possible for one of the condensation fluxes to become negative. This would imply a negative condensate composition by eqn. (2). Clearly there is a lower limit to the acceptable values of χ, and account must be taken of this in devising a robust computer algorithm. Having converged $\sum \tilde{y}_{iI}$ in successive passes of the inner iteration loop, its value is compared to unity. The dimensionless temperature χ is adjusted until $\sum \tilde{y}_{iI} = 1 \cdot 0$ within acceptable precision, as described above.

The following algorithm has proved to be successful in practice. A considerable number of the steps involved repeat those of the previous algorithm of Section (4.2) and are not listed again.

Step 1. Outer loop begins.
Guess χ. (defined by eqn. (64))

Step 2. Set : $\tilde{y}_{il} = \tilde{y}_{ig}$.
Either: $\tilde{x}_{il}$ is defined from prior condensation.
Or : $\tilde{x}_{il} = \tilde{x}^*_{ig}$ at vapour inlet (dew point calculation).

Steps 3–5. As in previous algorithm.

N.B. If location is the vapour inlet $T_l = T_l$; there is no iteration in Steps 4 and 5. Put α_l very large.

Step 6. Inner loop begins.

Steps 6–10. As in previous algorithm.

Step 11. First pass of inner loop. Estimate $K_i = K_i(T, P)$.
Later passes of inner loop. Estimate $K_i = K_i(T, P, \tilde{x}_{il}, \tilde{y}^*_{il})$.
ED: Calculate $\tilde{x}_{il}$, $i = 1, n - 1$ (eqn. (70)).
KS and LT: Calculate $\tilde{x}_{il}$, $i = 1, n - 1$ (eqn. (71)).
All: Calculate $\tilde{x}_{nl}$ (eqn. (72)).
Check: If $\tilde{x}_{nl}$ is negative χ has been guessed too small. Return to Step 4 using last successful χ value and thereafter decrease χ by smaller steps.

Step 12. From $\tilde{x}_{il}$ calculate $\tilde{y}^*_{il}$ with $\tilde{\theta} = 0$.
(Bubble point calculation) (eqns. (48, 49, 51)).
Calculate vapour film properties at $T_f(\tilde{y}^*_{il} + \tilde{y}_{ig})/2{\cdot}0$.
Calculate $\Delta\tilde{h}_v$ and $\tilde{c}_p$ (eqns. (66) and (2)).

Step 13. Calculate $\tilde{y}_{il}, \sum \tilde{y}_{il}$ (eqn. (51)).
Diagnosis: If successive estimates of $\sum \tilde{y}_{il}$ differ by more than the larger of 10^{-5} and $(\sum \tilde{y}_{il} - 1{\cdot}0)/10{\cdot}0$ the inner loop is *not converged*. Repeat from Step 7.

Step 14. Calculate $(\sum \tilde{y}_{il} - 1{\cdot}0)$.
Diagnosis: If $(\sum \tilde{y}_{il} - 1{\cdot}0)$ is greater than 10^{-5}, the outer loop is *not converged*. The value of χ is reset and T_f calculated (eqn. (68)). Calculation is repeated from Step 4.
Reset χ: (i) If the root of $(\sum \tilde{y}_i - 1{\cdot}0)$ has not been straddled continue to reduce χ by steps of 0·2 *unless* $\tilde{x}_{nl}$ has been negative in Step 11, when the reduction in χ will be smaller.
(ii) If the root $(\sum \tilde{y}_i - 1{\cdot}0)$ has been straddled, linearly interpolate from the last two values of χ to approximate the required root.

Step 15. Evaluate $\dot{q}_l$ (eqns. (1, 3)).

It is interesting to note that convergence of the vapour–liquid equilibrium calculations, which take place in Step 11, is achieved at the same time as convergence of physical properties and the matrix mass transfer methods within the inner loop.

The estimation of most physical properties is required within the inner loop and this has proved to make up a very significant fraction of the total computational time. Insufficient information is available on the sensitivity of the method to physical properties and the error that might be incurred by assuming them to be constant. A difficulty with the evaluation of physical properties is that the interfacial composition is not defined until convergence. This is overcome in the algorithm suggested by using the composition $\tilde{y}_{iI}^*$ in the evaluation of physical properties.

It has been previously suggested that a multidimensional Newton–Raphson procedure should be used in the determination of interfacial conditions for this case, Krishna *et al.*[5] In that approach $(2n + 1)$ evaluations of the condensation fluxes must be made by eqn. (9) in each stage of an iterative calculation of interfacial conditions. Typically, five stages are required giving a total number of evaluations of eqn. (9) of about $(10n - 15)$. In the method proposed above, convergence is usually achieved within about 25 passes of the inner loop and this should not change with the number of components. There is a clear advantage in using the above method for all cases of multicomponent condensation, for $n \geq 3$. However it becomes particularly useful with a large n.

4.2.3. Equations of Downstream Development

In design of a condenser handling an n component mixture, there are $(n + 1)$ independent variables which are functions of the cumulative heat exchanger area. These are vapour temperature T_g, coolant temperature T_c, and $(n - 1)$ independent vapour compositions, $\tilde{y}_{ig}$. By mass and energy balances over a differential element, the differential equations which describe how T_g, T_c and $\tilde{y}_{ig}$ vary with area A are:

$$\dot{N}_g \tilde{c}_{pg} \frac{dT_g}{dA} = -\alpha_g^{\cdot}(T_g - T_I) = -\alpha_g \frac{\varepsilon}{(e^{\varepsilon} - 1)}(T_g - T_I) \tag{73}$$

$$\dot{M}_c c_{pc} \frac{dT_c}{dA} = \pm \alpha_{oI}(T_I - T_c) \quad \begin{array}{l}(+ \text{ for cocurrent flow})\\(- \text{ for countercurrent flow})\end{array} \tag{74}$$

$$\frac{d}{dA}(\tilde{y}_{ig}\dot{N}_g) = -\dot{n}_i \qquad i = 1, n - 1 \tag{75}$$

This is an initial value problem and suitable boundary conditions are the specification of the values of the $(n + 1)$ independent variables at the vapour inlet $(T_g)_{in}$, $(\tilde{y}_{ig})_{in}$, and either $(T_c)_{out}$ for countercurrent flow or $(T_c)_{in}$ for cocurrent flow.

The condensate composition $\tilde{x}_{il}$, and total condensate flowrate $\dot{N}_l$, are obtained by an overall mass balance between the vapour inlet to the condenser and the position under consideration. Thus,

$$\dot{N}_l = (\dot{N}_g)_{in} - \dot{N}_g \tag{76}$$

$$\tilde{x}_{il} = \frac{(\tilde{y}_{ig}\dot{N}_g)_{in} - \tilde{y}_{ig}\dot{N}_g}{(\dot{N}_g)_{in} - \dot{N}_g} \tag{77}$$

In the case of countercurrent flow, two-point boundary conditions are available with vapour conditions known at the opposite end of the condenser to the coolant medium temperature. In this case a rough overall energy balance must be carried out to provide the coolant outlet temperature which may then be used to provide the appropriate boundary conditions for the initial value problem posed by eqns. (73)–(75).

The integration of eqns. (73)–(75) may be obtained by suitable integration methods. It is beyond the scope of this work to review such methods beyond stating that the Runge–Kutta–Merson method has proved successful.

Example 2. The condenser specified in Example 1 is to be reconsidered for the same duty; design to be carried out by a film model. It has been shown that in this case the effective diffusivity method introduced in Section (3.1) gives close agreement with both the matrix method and the linearised theory and should therefore be used in the calculation of local transfer rates. Propane is to be taken as the reference component n. The liquid condensate is to be assumed unmixed. The equations of downstream development are to be integrated using the Euler method with the condenser area divided into equal increments.

Solution. Local rates of heat and mass transfer are calculated by following the procedure outlined in detail in Section (4.2.2) with the effective diffusivity method (ED) used. The equations of Section (4.2.3) are integrated to show how conditions develop downstream. The full solution is summarised in Tables (4.4–4.6).

Table (4.4) shows this solution at a position one fifth of the way through the condenser for variables converged in the inner loop. It may be verified that vapour–liquid equilibrium is satisfied (eqns. (48–51) with $\tilde{\theta} = 0{\cdot}0$). It should also be noted that the unmixed liquid condition (eqn. (2)), and the equations describing the vapour phase diffusional process (eqns. (6) and (10)), which are used to derive eqn. (70) are all satisfied.

The converged solution of the outer loop of the procedure in Section (4.2.2) is shown in Tables (4.5 and 4.6), for the vapour inlet and five

TABLE 4.4

SOLUTION OF LOCAL EQUATIONS AT POSITION 3

Component	*Bulk vapour* $\tilde{y}_{ig}$	*Interfacial vapour* $\tilde{y}_{iI}$	*Interfacial condensate* $\tilde{x}_{iI}$	'K *values*' $K_i = \frac{\tilde{y}_{iI}}{\tilde{x}_{iI}}$	*Mean condensate* $\tilde{x}_{il}$	*Diffusive flux* $j_{ig} \times 10^4$ (kmol/m² s)	*Condensation flux* $\dot{n}_i \times 10^4$ (kmol/m² s)	$\frac{\beta^{\cdot}_{gin}}{\dot{n}_T}$
n-Octane C_8	0·190 53	0·028 56	0·334 91	0·085 28	0·326 33	0·726 56	1·685 38	0·891 39
n-Heptane C_7	0·236 04	0·077 51	0·390 08	0·198 71	0·382 94	0·775 16	1·963 00	0·971 66
n-Hexane C_6	0·042 80	0·026 80	0·059 76	0·448 42	0·059 24	0·085 33	0·300 73	1·059 32
n-Butane C_4	0·318 90	0·439 16	0·156 29	2·809 86	0·161 35	−0·818 30	0·786 53	1·352 11
n-Propane C_3	0·211 72	0·427 96	0·058 96	7·258 38	0·070 14	−0·768 75	0·296 71	—
Totals	1·000 0	1·000 0	1·000 0	—	1·000 0	0·000 0	5·032	—

TABLE 4.5
PHYSICAL PROPERTIES IN EXAMPLE 2

Property	*Position 1* (*inlet*)	*Position 2* $A_T/10{\cdot}0$	*Position 3* $A_T/5$	*Position 4* $2A_T/5$	*Position 5* $3A_T/5$	*Position 6* $4A_T/5$
Temperatures (K)						
T_{dew}	413·06	408·27	403·34	392·53	381·20	370·00
T_g	413·06	410·66	407·37	398·03	386·28	372·70
T_f	379·67	388·09	381·55	367·56	355·01	343·86
T_I	346·28	365·52	355·72	337·10	323·73	315·02
T_l	346·28	341·24	334·58	322·76	314·48	309·15
T_w	346·28	316·98	313·44	308·42	305·23	303·28
T_c	300·00	300·00	300·00	300·00	300·00	300·00
χ	0·409 31	0·592 09	0·518 95	0·378 50	0·274 98	0·206 58
Liquid properties (at T_l, $\tilde{X}_{il}$)						
Density (kg/m^3)	—	577·7	605·6	646·2	671·8	686·8
Specific heat capacity (J/kg K)	—	1 889·0	1 821·0	1 726·0	1 669·0	1 635·0
Viscosity ($Ns/m^2 \times 10^3$)	—	0·233 9	0·252 6	0·280 0	0·297 4	0·307 6
Thermal conductivity (W/m K)	—	0·121 6	0·122 5	0·124 2	0·125 4	0·126 1
Molecular Weight (kg/kmol)	—	92·02	93·23	93·15	92·25	91·31
Vapour film properties at $(T_f, (\tilde{y}_{il} + \tilde{y}_{ig})/2)$						
Density (kg/m^3)	8·110	7·989	7·769	7·457	7·273	7·182
Specific heat capacity (J/kg K)	2 087·0	2 118·0	2 097·0	2 049·0	2 005·0	1 964·0
Viscosity ($Ns/m^2 \times 10^5$)	0·891	0·907	0·907	0·900	0·890	0·877
Thermal conductivity (W/m K × 10)	0·162 3	0·165 8	0·166 4	0·152 6	0·148 7	0·144 3
Molecular Weight (kg/kmol)	69·63	70·32	67·36	62·41	58·82	56·23
Bulk vapour properties at $(T_g, \tilde{y}_{ig})$						
Density (kg/m^3)	9·075	8·677	8·349	7·786	7·375	7·134
Specific heat capacity (J/kg K)	2 201·0	2 194·0	2 185·0	2 156·0	2 118·0	2 071·0
Viscosity ($Ns/m^2 \times 10^5$)	0·897	0·908	0·917	0·927	0·929	0·920
Thermal conductivity (W/m K × 10)	0·155 1	0·161 1	0·165 4	0·171 4	0·173 6	0·171 7
Molecular Weight (kg/kmol)	84·07	80·32	76·98	70·57	65·11	60·84
Interface properties at $(T_I, \tilde{y}_{il}, \tilde{x}_{il})$						
Latent heat (MJ/kmol)	33·01	34·48	33·72	31·82	29·97	28·46

TABLE 4.6

DESIGN CALCULATIONS IN EXAMPLE 2

	Position 1 (inlet)	Position 2 $A_T/10{\cdot}0$	Position 3 $A_T/5$	Position 4 $2A_T/5$	Position 5 $3A_T/5$	Position 6 $4A_T/5$
Flowrates per tube						
Vapour flowrates (kmol/s)						
C_8 ($\times 10^4$)	0·270 6	0·162 2	0·116 3	0·060 1	0·031 7	0·017 4
C_7 ($\times 10^4$)	0·325 1	0·195 0	0·144 0	0·077 5	0·041 7	0·023 0
C_6 ($\times 10^4$)	0·054 1	0·033 3	0·026 1	0·015 6	0·009 1	0·005 3
C_4 ($\times 10^4$)	0·270 9	0·210 2	0·194 5	0·164 2	0·135 8	0·112 1
C_3 ($\times 10^4$)	0·162 4	0·135 3	0·129 2	0·117 9	0·107 8	0·099 4
Total $\dot{M}_g$	0·009 105	0·005 912	0·004 697	0·003 072	0·002 124	0·001 564
Condensate flowrate $\dot{M}_l$ (kg/s)	0·0	0·003 193	0·004 408	0·006 033	0·006 981	0·007 541
Calculation of α_g						
Reynolds number Re_g	50 899	32 632	25 688	16 610	11 462	7 548
Chilton–Colburn J_H factor	0·003 64	0·003 93	0·004 09	0·004 41	0·004 70	0·005 04
Prandtl number Pr_f	1·146	1·154	1·211	1·166	1·133	1·110
Gas film h.t.c. α_g (W/m² K)	124·81	88·07	72·82	48·27	34·95	26·61
Condensate h.t.c. α_l (W/m² K)	—	594·1	540·8	498·1	481·7	474·0
Total flux $\dot{n}_T$ (kmol/m² s)	0·001 784	0·000 646 8	0·000 503 2	0·000 313 1	0·000 197 2	0·000 126 3
ε	2·077	1·555	0·976 1	0·829 2	0·664 8	0·522 9
Mass transfer coefficients						
β_{1n} (kmol/m² s) $\times 10^4$	11·55	8·064	6·689	4·739	3·509	2·732
β_{2n}	12·28	8·573	7·112	5·038	3·731	2·905
β_{3n}	13·07	9·125	7·570	5·365	3·972	3·093
β_{4n}	15·69	10·96	9·090	6·442	4·770	3·714
Total heat flux $\dot{q}_t$ (W/m²)	78 671	28 843	22 860	14 294	8 905	5 566
$\dot{q}_I$ (W/m²)	19 796	6 539	5 891	4 327	2 994	1 971

downstream positions. The various important temperatures and all physical properties evaluated under appropriate conditions are given in Table (4.5), while the calculation of heat transfer coefficients is summarised in Table (4.6). The temperatures satisfy eqns. (64, 68 and 69) while the heat transfer coefficients satisfy eqns. (24 and 44) (with $b_1 = 0{\cdot}023$, $b_2 = -0{\cdot}17$, $b_3 = 0{\cdot}667$). The Chilton–Colburn analogy between heat and mass transfer has been used and the β_{gin} are evaluated by eqn. (45). Local mass and energy fluxes, the objective of the local calculations, are also given in Table (4.6). It is readily verified that the total flux of condensation satisfies eqns. (65–7) and the energy fluxes in gas and liquid films satisfy eqns. (1, 3, 5 and 8).

The variation of gas temperature (Table (4.5)) and vapour flowrates (Table (4.6)) have been predicted by Euler's method in which the derivatives in eqns. (73 and 75) are replaced by differences,

$$\dot{N}_g \tilde{c}_{pg} \Delta T_g = -\alpha_g \varepsilon (T_g - T_I) \Delta A / (\exp(\varepsilon) - 1) \tag{73}$$

$$\Delta(\dot{N}_{ig}) = \Delta(\tilde{y}_{ig} \dot{N}_g) = -\dot{n}_i \qquad i = 1, n-1 \tag{75}$$

The condenser area has been divided into 10 equal increments and these difference equations applied in each interval. It is readily verified that the results reported at positions 1–3 (at intervals of $A_T/10$) satisfy these equations.

It is not suggested that the Euler method provides a sufficiently accurate design for practical purposes but has been presented here to provide a basis for the development of computational procedures, and because the results may be verified by hand calculation. The design presented is *not* conservative in that fluxes calculated at the beginning of an interval, where heat transfer coefficients are highest, are used to predict condensation rates over the entire interval.

Table (4.7) compares important outlet conditions for the Kern and Silver methods and the film model, described above. The film model shows very slight subcooling of the vapour at the outlet but from Table (4.5) it is seen to be superheated at intermediate positions. However departures from

TABLE 4.7

Outlet condition	*Kern*	*Silver*	*Film*
Vapour temperature (K), T_g	321·6	356·0	358·8
Saturation temperature of vapour	321·6	356·0	359·3
Fraction condensed (mole %), $100\,(1-\tilde{\theta})$	100·0	72·4	80·0
Heat load (MW) $\dot{Q}_T$	6·00	4·43	4·21

saturation are small and the assumption of equilibrium in the approximate methods is reasonable. The Silver method and film model are in reasonable agreement on the fraction condensed, heat load and outlet temperature. However in the film model the bubble point of a liquid of the outlet vapour composition is less than 300 K and complete condensation is impossible. This has arisen because the liquid is assumed to be unmixed. The heat load predicted by the film model is about 10 % smaller than would be expected, but this is due to neglect of liquid subcooling, which is partially accounted for in the equilibrium methods.

The absence of data on large condensers prevents positive recommendations as to the reliability of the above methods. Particularly serious from the point of view of the above example is the absence of information on liquid mixing. It is clear however that the film model is a closer representation of physical reality and with further development is capable ultimately of better reliability.

5. THE DESIGN OF CONDENSERS OF PRACTICAL GEOMETRIES

5.1. Baffled Shell-and-tube Condensers

The shell-and-tube heat exchanger is readily adaptable to condensing service. Thermal design, however, is considerably more difficult than in the single-phase applications. As we have remarked, a stepwise calculation procedure is generally necessary and in addition local transfer rates must be calculated iteratively. Calculation of the transport and thermodynamic properties, and the treatment of the vapour–liquid equilibrium in multicomponent condensation also require considerable computational effort. Consequently there is a strong incentive to computerise the design procedure, and organisations such as HTFS† and HTRI‡ have developed sophisticated digital computer programs for condenser simulation and design. Certainly the stepwise procedures described earlier in this chapter may only be used in computer design and simulations.

Here we describe techniques, applicable to computer calculations, for integration of the local information on heat transfer rates described previously over the whole exchanger surface area to yield the thermal design in cases where the geometry is more complex than in Section (4).

† HTFS Heat Transfer and Fluid Flow Service, Harwell and NEL, UK.
‡ HTRI Heat Transfer Research Incorporated, Los Angeles, California, USA.

Techniques for the equally important pressure drop calculations are also described.

In the design of baffled shell-and-tube heat exchangers it is usually assumed that only axial variations in heat transfer rate are significant. This renders the methods unsuitable for cross-flow condensers.

Heat Transfer Calculations

A considerable simplification in the overall heat transfer calculation is possible if temperature–enthalpy relationships are available for each fluid stream at the start of the calculation. We first describe methods of this type, which are therefore applicable when the Silver[50] type of method is used to evaluate local coefficients.

Method Using Predetermined Temperature–Enthalpy Relationship

The main steps in the method are as follows, but more detail is included in Section (4.1).

(1) Calculation of condensing stream temperature–enthalpy curve. The temperature–enthalpy curve can be evaluated in one of two ways.

(a) Integral condensation curve. Here the vapour mixture at any position in the condensing range is considered to be in equilibrium with all of the accumulated condensate at the bulk vapour temperature. The temperature–enthalpy relationship can be derived as shown in Section (4.1) above, where $\tilde{h}$ is evaluated at a number of temperature points using eqn. (52).

(b) Differential condensation curve. The condensate is considered to drop out of contact with the remaining vapour.

Neither of these cases is realised, since the condensate forms at some temperature intermediate to the vapour and coolant temperatures whenever there is a gas-phase resistance. Differential condensation can result in failure to condense the lighter components within the limits of the process conditions. Integral condensation implies that the entire condensate flow is interposed between the vapour and tube wall, leading to a higher condensate resistance. The integral condensation curve is almost universally used, partly because it is much easier to calculate.

(2) Evaluation of mean temperature differences. The mean temperature difference in general varies non-linearly over the condensing range. It is appropriate within the framework of this method to divide the temperature–enthalpy curve into a number of zones in which both the shell-side and tube-side temperature–enthalpy curves can be approximated by

straight lines. Thus in a single-pass condenser the zone mean temperature ΔT_{mz} is given by eqn. (20)

$$\Delta T_{mz} = \{\Delta T_{in} - \Delta T_{out}\}/\ln(\Delta T_{in}/\Delta T_{out}) \tag{78}$$

where ΔT_{in} and ΔT_{out} refer to the ends of the zone.

When a condenser has two or more tube passes we encounter the difficulty that the local overall coefficients on each pass at any cross section are generally different. With two tube passes the equation corresponding to eqn. (15) is

$$d\dot{Q} = U^{I}\Delta T^{I}\frac{dA}{2} + U^{II}\Delta T^{II}\frac{dA}{2} \tag{79}$$

where the superscripts I and II refer to the tube passes. The problem is greatly simplified if it is assumed that U^{I} and U^{II} can be replaced by a mean U found from a local calculation of the mean condition. This approximation was made by Bell and Ghaly,[51] who suggest that the assumption is nearly always reasonable with shell-side condensation, though it is not valid for tube-side condensation. Bell and Ghaly then apply a heat balance over the area dA to obtain

$$-\frac{dT_g^{II}}{dT_g^{I}} = \frac{\Delta T^{II}}{\Delta T^{I}} \tag{80}$$

Butterworth[53] subsequently generalised the method by writing eqn. (80) in terms of enthalpies:

$$-\frac{d\tilde{h}_c^{II}}{d\tilde{h}_c^{I}} = \frac{\Delta T^{II}}{\Delta T^{I}} \tag{81}$$

This does not require uniform specific heat on the coolant side. The procedure given by Butterworth[53] is to integrate eqn. (81) along the condenser starting from the known values of $\tilde{h}_c$ at inlet and outlet and using the known temperature–enthalpy relationships. Equation (79) becomes

$$d\dot{Q} = U(T_g - \bar{T}_c)\,dA \tag{82}$$

where,

$$\bar{T}_c = \tfrac{1}{2}(T_c^{I} + T_c^{II}) \tag{83}$$

The mean temperature difference for each zone can now be evaluated using eqn. (78) with ΔT replaced by $(T_g - \bar{T}_c)$. This method can be applied to TEMA[54] E-shells with four tube-passes, and to TEMA J-shells. With more than four passes the temperature distribution calculated for four passes can be used with little error.

Emerson[55] has proposed a similar type of method in which TEMA E-type shells with four or more tube passes are considered to have a temperature distribution calculated analytically for an infinite number of tube passes.

(*3*) *Calculation of the surface area requirement*. As pointed out by Butterworth,[53] one extremely useful feature of the above methods is that the temperature difference calculation is independent of the heat transfer rate. For a given duty only one temperature distribution calculation is required. Various geometries can then be tested for the surface area requirement, which for each zone is given by

$$A_z = \frac{\dot{N}_g\{\tilde{h}_{gin} - \tilde{h}_{gout}\}}{U_{mz}\Delta T_{mz}} \tag{84}$$

and an optimised design obtained. The mean zone coefficient U_{mz} is given by

$$U_{mz} = \tfrac{1}{2}(U_{in} + U_{out}) \tag{85}$$

Further details of the procedure are given by Butterworth[53] and Emerson.[55]

Method for Film Models

When local condensation rates are evaluated using the film model approach described in Section (4.2), it is necessary to solve a set of differential equations describing the variation of T_g, T_c and $\tilde{y}_{ig}$ with area. This approach is suited to direct simulation calculations. The temperature and composition profiles are evaluated explicitly and there is no need for condensation curves. One decision that must be taken prior to the solution concerns the model adopted for the liquid phase concentration at the interface. As described previously there are two alternatives, perfect liquid mixing and no liquid mixing.

The perfect mixing and no mixing models are analogous to the integral and differential condensation curves respectively. As the example in Section (4.2) above shows, the design can be sensitive to the model chosen, particularly if there is no non-condensing gas present. Reality lies somewhere between the two extremes, but there is little experimental evidence available to guide the designer. While the no-mixing option seems more appropriate to horizontal shell-side condensation, the designer who is accustomed to using the integral condensation curve may prefer the perfect-mixing model.

For the simulation calculation the equations of downstream development are integrated numerically as described in Section (4.2). With multi-pass units the initial conditions at either end of the exchanger are incomplete, and a relaxation type of method is required.

Pressure drop calculation. Standard two-phase methods as given, for example, in Collier[56] can be applied to obtain the frictional pressure gradient during condensation in tubes. Since, in condensation, momentum changes lead to a recovery in pressure, it is possible to ignore the momentum term and count it as a safety factor. Several authors have also considered the effect of mass transfer on the shear stress at the vapour–liquid interface. Analytical treatments involve some assumption about the axial velocity of the condensing vapour, and there is some controversy in the literature on this point.

Much less information is available on pressure drop with shell-side condensation. Whilst reasonably sophisticated 'stream-analysis' models have been developed for single-phase flow, no such method is yet available for two-phase flow. There is also a general scarcity of two-phase pressure drop and flow pattern information. Grant and Chisholm[57] present flow pattern maps based on observations from a model heat exchanger with flow of air–water mixtures. They correlate their frictional pressure drop data using the equation

$$\frac{\Delta p_{TP}}{\Delta p_{lo}} = 1 + (\Gamma^2 - 1)\{Bx^{(2-n)/2}(1-x)^{(2-n)/2} + x^{2-n}\} \tag{86}$$

where Γ is a physical property coefficient defined by

$$\Gamma = \left(\frac{\Delta p_{go}}{\Delta p_{lo}}\right)^{1/2} \tag{87}$$

and n is the Blasius exponent for single-phase flow. For the crossflow zone the approximate values of B were for horizontal side-to-side spray and bubbly flow $B = 0{\cdot}75$, and for horizontal side-to-side stratified-spray flow $B = 0{\cdot}25$.

For the window zone, where $n = 0$, the value of B for side-to-side flow was obtained as

$$B = \frac{2}{\Gamma + 1}$$

5.2. Crossflow Condensers

Units which are designed such that the vapour flows over the tube bank in a

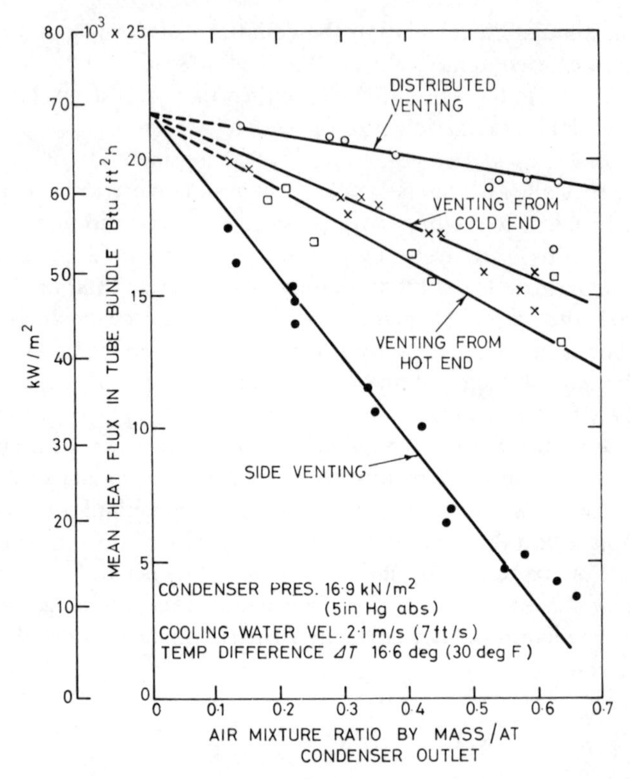

FIG. 4.5. Comparison of condenser performance with different venting arrangements (from ref. 62).

single pass are known as cross-flow condensers. This is in contrast to the baffled condenser in which the vapour flow is designed to turn and cross the bundle several times. Cross-flow condensers are used throughout the power and process industries, particularly where a low pressure drop is required. A common example is the surface condenser used in power plant. With the advent of 500 MW (and larger) generating sets, the modern power station condenser represents a very large undertaking indeed. A typical 500 MW unit has a surface area of around 28 000 m^2 with cooling water flows of 15 m^3/s. Design of these units is a specialised task involving the economic optimisation of many factors. For a review of recent design practice the reader is referred to papers by Christie,[58] Andrews[59] and Waterton.[60] Special considerations applying to marine and naval condensers are discussed by MacNair.[61] Here we confine ourselves to recent developments in the understanding of certain aspects of heat transfer and pressure change which apply to both power plant and process condensers.

Some important considerations in the thermal design of cross flow condensers are:

(1) Where a non-condensing gas is present, provision must be made for its continuous removal. Grant[62] has performed an extensive study of the influence of the location of the non-condensing gas vent on the performance of a steam condenser with 149 tubes of length 2·24 m. In separate tests the condenser was vented from the side of the bundle and from the centre of the tube nest using a sparge tube. With centre venting the air extraction points were located (a) uniformly over the whole length of the sparge tube, (b) at the coolant inlet end, and (c) at the coolant outlet end. The relative performance of the four methods is illustrated in Fig. 4.5, which shows that side venting gives the poorest performance. Some other examples of poorly vented condensers are discussed by Standiford.[63]

(2) In view of the low pressure drop requirement it is sometimes undesirable to force the entire vapour flow over the first rows of the condenser. By-pass lanes between the bundle and the shell outer wall, and open access lanes penetrating into the bundle are therefore provided. Emerson[64] reports tests on a rectangular condenser in which the shell wall can be moved away from the bundle, thus varying the width of the by-pass lane. With venting below the nest, Fig. 4.6 shows that with an exit air mass fraction of 0·3, increasing the by-pass width from 0 to 50 mm reduces the condenser performance by two-thirds. It was also found that in a bundle of roughly circular cross-section with centre venting, the size of the bundle–shell clearance had almost negligible effect.

(3) The temperature gradient of the coolant influences the axial condensation rate. With the relatively low temperature differences found in steam condensers, significant variations in heat flux between the coolant inlet and outlet ends can occur.

From the above it is apparent that some assessment of the fluid flow field over the tube nest is essential for good design. The cross-flow condenser in fact exhibits a strong inter-dependence between fluid flow and heat transfer. A realistic simulation requires a two- or three-dimensional solution of the forced convection and diffusional flow patterns across the tube nest cross-section.

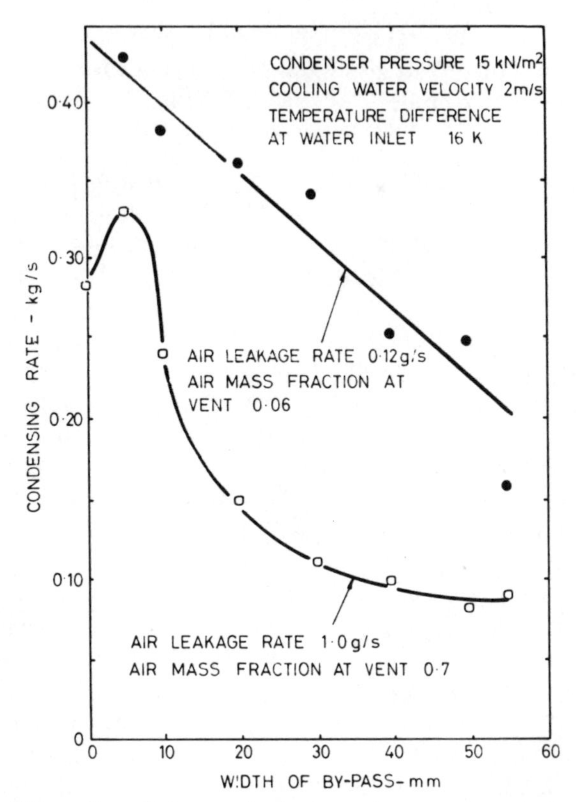

FIG. 4.6. Effect of small changes in the space between the tube bundle and the shell (from ref. 64).

Wilson[26] first produced a computer method for solution of the two-dimensional equations. Since the flow and heat transfer distributions are inter-related they must be solved for simultaneously. The procedure is to apply an arbitrary pressure distribution over a number of nodes in the tube field and from that calculate the mass flow and air distributions, and hence the local condensation rates. Iteration is performed on the pressure distribution until a mass balance at each node is achieved. In Wilson's solution only the pressure change due to friction is considered, and the vapour and gas are assumed to have the same velocity throughout. The program has successfully predicted the location of pockets of non-condensing gas.

Davidson[65] included the accelerational component of pressure change in his two-dimensional finite difference analysis of the continuity and momentum equations. Murray[66] has recently proposed a finite element technique, taking into account bulk diffusion of the gas through the vapour, which can be significant where bulk flow velocities are low.

The cost of running two-dimensional computer programs is considerable by normal heat exchanger design standards. For this reason these methods are applied only to power condensers, where the economic value of design improvements is significant, or to certain critical applications, such as marine condensers.

Despite the sophistication of the analysis, the two-dimensional methods are limited in the sense that many large condensers exhibit significant axial variations in heat transfer rate, and there is consequently bulk flow of vapour from the hot end to the cold end of the condenser. Further, the necessary information on the variation of heat transfer coefficients and pressure loss coefficients with vapour flow, flow direction and condensate loading is largely incomplete. Nevertheless two-dimensional methods represent an extremely powerful analytical tool.

5.3. Direct Contact Condensers

Direct contact condensers can be used where it is permissible to mix the vapour and coolant. They offer relatively cheap construction and a high heat transfer pressure drop ratio. Fouling problems are absent, and the direct contact condenser has a more flexible range of operation than the shell-and-tube condenser. The most common type is the spray condenser, and baffle columns and packed columns are also used in certain applications. An alternative is to bubble the vapour through the coolant.

There is no established scientific basis for the design of direct contact condensers. Wherever possible plant data should be used and Fair[67] supplied empirical correlations of heat transfer and pressure drop in spray condensers and tray columns. However these correlations should not be applied where the process conditions are outside the range of the data, or to different geometries. It is quite possible to formulate design procedures from theory, but the difficulty is that many of the quantities required are not known to an acceptable degree of accuracy. There are several uncertainties, particularly if a non-condensing gas is present. Modelling of the flow patterns (often recirculating) in spray chambers is difficult, and, in the usual vacuum application, interfacial resistance can be significant.

Recent research has been directed at certain specialised applications. Jacobs *et al.*,[68] motivated by the possible economic advantages of direct contact binary condensers in geothermal power plants have studied condensation of a vapour on an immiscible liquid in a packed column, and have correlated their own and other data for gravity controlled flows. Bakay and Jaszay[69] describe a high performance jet condenser for medium size power plant applications.

An important application of the direct contact condenser is in the pressure suppression systems of water-cooled nuclear reactors. Here steam is injected downwards into a pool of subcooled water. Marks and Andeen[70] and Sargis *et al.*[71] have examined the potentially damaging phenomena of steam 'chugging' and condensation oscillation which can occur in these systems. An earlier study is that of Rumary *et al.*[72] who refer to the same phenomena as 'pumping' and 'cracking'. Other recent references are Aoki *et al.*,[73] Cumo *et al.*[74] and Chan.[75]

REFERENCES

1. TANASAWA, I. Dropwise condensation—the way to practical applications, Sixth International Heat Transfer Conference, 1978, **6,** 393–405.
2. DEO, P. V. PhD Thesis, 1979, University of Manchester.
3. SARDESAI, R. G. PhD Thesis, 1979, University of Manchester.
4. SCHRAGE, R. W. *Interphase Mass Transfer*, 1953, Columbia University Press, New York.
5. KRISHNA, R., PANCHAL, C. B., WEBB, D. R. and COWARD, I. An Ackermann–Colburn and Drew type analysis for condensation of multicomponent mixtures, *Lett. Heat Mass Trans.*, 1976, **3,** 163–72.
6. ACKERMANN, G. *V.D.I. Forschungsheft*, 1937, **382,** 1.
7. COLBURN, A. P. and DREW, T. B. The condensation of mixed vapours, *Trans. Am. Inst. Chem. Engnrs.*, 1937, **33,** 197.
8. TOOR, H. L. Solution of the linearised equations of multicomponent mass transfer, *A.I.Ch.E.J.*, 1964, **10**(4), 448, 460.
9. STEWART, R. E. and PROBER, R. Matrix calculation of multicomponent mass transfer in isothermal systems, *Ind. Eng. Chem. Fundamentals*, 1964, **3,** 224.
10. KRISHNA, R. and STANDART, G. L. A multicomponent film model incorporating a general matrix method of solution to the Maxwell–Stefan equations, *A.I.Ch.E.J.*, 1976, **22**(2), 383.
11. WEBB, D. R. and PANCHAL, C. B. The significances of multicomponent diffusional interactions in the process of condensation is the presence of a non-condensable gas, 1979, HTFS Research Symposium, Paper RS295.
12. TAYLOR, R. and WEBB, D. R. Stability of the film model of multicomponent mass transfer, *Chem. Eng. Comm.*, 1980 (in press).
13. KRISHNA, R. and STANDART, G. L. Mass and energy transfer in multicomponent systems, *Chem. Eng. Comm.*, 1979, **3,** 201–75.
14. NUSSELT, W. The surface condensation of water vapour, *V.D.I. Z.*, 1916, **60,** 541–46, 569–75.
15. JASTER, H. and KOSKY, P. G. Condensation heat transfer in a mixed flow regime, *Int. J. Heat Trans.*, 1976, **19,** 95–99.
16. BREBER, G., PALEN, J. W. and TABOREK, J. Prediction of horizontal tubeside

condensation of pure components using flow regime criteria, in: *Condensation Heat Transfer*, 1979, A.S.M.E., New York, 1–8.

17. TAITEL, Y. and DUKLER, A. E. A model for predicting flow regime transitions in horizontal and near horizontal gas–liquid flow, *A.I.Ch.E. J.*, 1976, **22**(1) 47–55.
18. PALEN, J. W., BREBER, G. and TABOREK, J. Prediction of flow regimes in horizontal tubeside condensation, 17th National Heat Transfer Conference, 1977, Salt Lake City, A.I.Ch.E. Paper No. 5.
19. ZIVI, S. M. Estimation of steady state steam void fraction by means of the principle of minimum entropy production, *J. Heat Trans.*, 1964, **86,** 247–52.
20. BOYKO, L. P. and KRUZHILIN, G. N. Heat transfer and hydraulic resistance during condensation of steam in a horizontal tube and a bundle of tubes, *Int. J. Heat Mass Trans.*, 1967, **10,** 361–73.
21. KOSKY, P. G. and STAUB, F. W. Local condensation heat transfer coefficients in the annular flow regime, 1971, *A.I.Ch.E. J.*, **17**(5), 1037–43.
22. ALTMAN, M., STAUB, F. W. and NORRIS, R. H. Local heat transfer and pressure drop for refrigerant-22 condensing in horizontal tubes, *Chem. Eng. Prog. Symp. Ser.*, 1960, **56**(30), 151–59.
23. TRAVISS, D. P., BARON, A. B. and ROHSENOW, W. M. Forced convective condensation inside tubes, 1971, MIT Report number DSR 72591-74.
24. BAE, S., MAULTBETSCH, J. S. and ROHSENOW, W. M. Refrigerant forced-convection condensation inside horizontal tubes, 1970, MIT Report number DSR 72591-71.
25. CARPENTER, E. F. and COLBURN, A. P. The effect of vapour velocity on condensation inside tubes, 1951, Trans. I. Mech. E. and A.S.M.E., General Discussion on Heat Transfer, 20–26.
26. WILSON, J. L. The design of condensers by digital computers, *Inst. Chem. Eng. Symp. Ser.*, 1972, **35,** 3:21–3:27.
27. GRANT, I. D. R. and OSMENT, B. D. J. The effect of condensate drainage on condenser performance, 1968, N.E.L. Report No. 350, East Kilbride, Glasgow National Engineering Laboratory.
28. SHEKRILADZE, I. G. and GOMELAURI, V. I. Theoretical study of laminar film condensation of flowing vapour, *Int. J. Heat Mass Trans.*, 1966, **15,** 235–46.
29. FUJII, T., UEHARA, H. and KURATA, C. Laminar filmwise condensation of flowing vapour on a horizontal cylinder, *Int. J. Heat Mass Trans.*, 1972, **15,** 235–46.
30. NICOL, A. A. and WALLACE, D. J. Condensation with appreciable vapour velocity and variable wall-temperature, in: *Steam Turbine Condensers*, 1976, Report of a Meeting at N.E.L., 17–18 September 1974, N.E.L. Report No. 619, East Kilbride, Glasgow, National Engineering Laboratory.
31. NICOL, A. A., BRYCE, A. and AHMED, A. S. A. Condensation of a horizontally flowing vapour on a horizontal cylinder normal to the vapour stream, Sixth International Heat Transfer Conference, 1978, **2,** Paper CS-4, 401–6.
32. NOBBS, D. W. and MAYHEW, Y. R. Effect of downward vapour velocity and inundation on condensation rates on horizontal tube banks, 1976, N.E.L. Report No. 619, East Kilbride, Glasgow, National Engineering Laboratory.
33. FUJII, T., HONDA, H. and ODA, K. Condensation of steam on a horizontal tube —the influence of oncoming velocity and thermal condition at the tube wall, in: *Condensation Heat Transfer*, 1979, A.S.M.E., New York.

34. Bernhardt, S. H., Sheridan, J. J. and Westwater, J. W. Condensation of immiscible mixtures, *A.I.Ch.E. Symp. Ser.*, 1972, **68**(118), 21–37.
35. Akers, W. W. and Turner, M. M., Condensation of vapours of immiscible liquids, *A.I.Ch.E. J.*, 1962, **8**, 587–89.
36. Polley, G. T. and Calus, W. F. The effect of condensate pattern on heat transfer during the condensation of binary mixtures of vapours of immiscible liquids, Sixth International Heat Transfer Conference, 1978, **2**, Paper CS-16, Toronto.
37. Chilton, T. H. and Colburn, A. P. Mass transfer coefficients. Prediction from data on heat transfer and fluid friction, *Ind. Eng. Chem.*, 1934, **26**, 1183–87.
38. Price, B. C. and Bell, K. J. Design of binary vapour condensers using the Colburn–Drew equations, *A.I.Ch.E. Symp. Ser.*, 1974, **70**(138), 163–71.
39. Bird, R. B., Stewart, W. E. and Lightfoot, E. N. *Transport Phenomena*, 1960, Wiley, New York.
40. Berman, L. D. Determining the coefficients of heat and mass transfer in calculations of vapour condensation from vapour–gas mixtures, *Therm. Eng.*, 1972, **19**(11), 72–76.
41. Berman, L. D. On the effect of molecular-kinetic resistance upon heat transfer with condensation, *Int. J. Heat Mass Trans.*, 1967, **10**, 1463.
42. Chao, K. C. and Seader, J. D. A general correlation of vapour–liquid equilibria in hydrocarbon mixtures, *A.I.Ch.E. J.*, 1961, **7**(4), 598.
43. Cavett, R. H. Physical data for distillation calculations—vapour–liquid equilibria, *Proc. Am. Petr. Inst.*, 1962, **42**(3), 351.
44. Grayson, H. G. and Streed, C. W. Vapour–liquid equilibria for high temperature, high pressure hydrogen–hydrocarbon systems, Sixth World Petroleum Congress, 1963, Section VII, Paper 20-PD7.
45. Shelton, R. J. A computer program for calculating flash equilibrium characteristics and heat contents of hydrocarbon systems, *The Chemical Engineer*, 1968, (224), 385.
46. API Technical Data Book—Petroleum Refining, 1971, American Petroleum Institute, New York.
47. Edmister, W. C. Applied hydrocarbon thermodynamics, Part 41—partial enthalpies from Redlich–Kwong equation, 1971, *Hydrocarbon Proc.*, **50**(9), 183.
48. Reid, R. C. and Sherwood, T. K. *The Properties of Gases and Liquids*, 1966, McGraw-Hill, New York.
49. Kern, D. Q. *Process Heat Transfer*, 1950, McGraw-Hill, New York.
50. Silver, L. Gas cooling with aqueous condensation, *Trans. Inst. Chem. Eng.*, 1947, **25**, 30–42.
51. Bell, K. J. and Ghaly, M. A. An approximate generalized design method for multicomponent/partial condensers, *A.I.Ch.E. Symp. Ser.*, 1972, **69**(131), 72–79.
52. Ward, D. J. How to design a multiple component partial condenser, *Petrol. Eng.*, 1960, **32**(10), C42.
53. Butterworth, D. A calculation method for shell and tube heat exchangers in which the overall coefficient varies along the length, in: *Advances in Thermal and Mechanical Design of Shell-and-Tube Heat Exchangers*, 1975, Report of a Meeting at N.E.L., 28 November 1973, N.E.L. Report No. 590, East Kilbride, Glasgow, National Engineering Laboratory.

54. TEMA *Standards of the Tubular Exchanger Manufacturers' Association*, 1978, (6th edition), NY: Tubular Exchanger Manufacturers Association, New York.
55. EMERSON, W. H. Effective tube-side temperatures in multi-pass heat exchangers with non-uniform heat-transfer coefficients and specific heats, in: Advances in Thermal and Mechanical Design of Shell-and-Tube Heat Exchangers, 1975, Report of a meeting at N.E.L., 28 November 1973, N.E.L. Report No. 590, East Kilbride, Glasgow, National Engineering Laboratory.
56. COLLIER, J. G. *Convective Boiling and Condensation*, 1972, McGraw-Hill, New York.
57. GRANT, I. D. R. and CHISHOLM, D. Two-phase flow on the shell-side of a segmentally baffled shell-and-tube heat exchanger, *J. Heat Trans.*, 1979, **101,** 36–42.
58. CHRISTIE, D. G. Design of condensers for large turbo-generators, in: *Steam Turbine Condensers*, 1976, Report of a meeting at N.E.L., 17–18 September 1974, N.E.L. Report No. 619, East Kilbride, Glasgow, National Engineering Laboratory.
59. ANDREWS, F. C. A review of the thermal performance achievements of 500 MW condensing plant—some aspects of steam-side venting, N.E.L. Report No. 619, 1976, East Kilbride, Glasgow, National Engineering Laboratory.
60. WATERTON, J. C. Aspects of condenser-design, N.E.L. Report No. 619, 1976, East Kilbride, Glasgow.
61. MACNAIR, E. J. Introductory remarks on marine and naval condensers, 1976, N.E.L. Report No. 619, East Kilbride, Glasgow, National Engineering Laboratory.
62. GRANT, I. D. R. Condenser performance—the effect of different arrangements for venting non-condensing gases, *Brit. Chem. Eng.*, 1969, **14**(12), 1709–11.
63. STANDIFORD, F. C. Effect of non-condensables on condenser design and heat transfer, 1979, 18th National Heat Transfer Conference, San Diego.
64. EMERSON, W. H. Introductory survey, 1976, N.E.L. Report No. 619, East Kilbride, Glasgow, National Engineering Laboratory.
65. DAVIDSON, B. J. Computational methods for evaluating the performance of condensers, 1976, N.E.L. Report No. 619, East Kilbride, Glasgow, National Engineering Laboratory.
66. MURRAY, W. L. Numerical solutions of the diffusion–convection equation in steam condenser flow modelling, Paper presented at International Conference on Numerical Methods in Thermal Problems, 1979, University of Wales, Swansea.
67. FAIR, J. R. Designing direct contact coolers/condensers, *Chem. Eng.*, 1972, **79,** 91–100.
68. JACOBS, H. R., THOMAS, K. D. and BOEHM, R. F. Direct contact condensation of immiscible fluids in packed beds, in: *Condensation Heat Transfer*, 1979, A.S.M.E., New York, 103–10.
69. BAKAY, A. and JASZAY, T. High performance jet condensers for steam turbines, 1978, Sixth International Heat Transfer Conference, Paper EC-10, 261–65.
70. MARKS, J. S. and ANDEEN, G. B. Chugging and condensation oscillation, in: *Condensation Heat Transfer*, 1979, A.S.M.E., New York, 93–101.
71. SARGIS, D. A., MASIELLO, P. J. and STUHMILLER, J. H. A probabilistic model for predicting steam chugging phenomena, in: *Condensation Heat Transfer*, 1979, A.S.M.E., New York.

72. Rumary, C. H., Smith, I. J. and Smith, M. J. S. The efficiency of a water pond for the direct condensation of steam–air mixtures, 1970, *Direct Contact Heat Transfer:* Report of a Meeting at N.E.L., 15 January 1969. N.E.L. Report No. 453, East Kilbride, Glasgow, National Engineering Laboratory.
73. Aoki, S., Inoue, A., Kozawa, Y. and Akimoto, H. Direct contact condensation of flowing steam onto injected water, 1978, Sixth International Heat Transfer Conference, Paper NR-19, **5,** 107–12.
74. Cumo, M., Farello, G. E. and Ferrari, G. Direct heat transfer in pressure-suppression systems, Sixth International Heat Transfer Conference, 1978, **5,** Paper NR-18, 101–6.
75. Chan, C. K. Dynamical pressure pulse in steam jet condensation, 1978, Sixth International Heat Transfer Conference, **2,** Paper ICS-18, 395–99.

Chapter 5

COMPACT HEAT EXCHANGERS

J. D. Usher and G. S. Cattell
The A.P.V. Company Ltd, Crawley, UK

SUMMARY

The term 'compact heat exchanger' is capable of a number of different interpretations. For the purposes of the chapter, a rigid lower limit of surface area/volume ratio is not employed. Instead, a more general criterion, defining those configurations capable of providing a surface area/volume ratio significantly greater than an orthodox shell-and-tube unit was adopted.

The means by which compactness is achieved has been divided into two areas; the need for an extended heat-transfer surface as in gas-flow applications, and the use of plates in place of tubular surfaces as in the gasketted plate unit, the lamella and spiral heat exchangers. In addition, the carbon block unit is included as it derives compactness from close spacing of flow passages within the block itself.

For each of the configurations described, the mechanical layout, its advantages and limitations are discussed, providing typical pressure and temperature resistance, area and flowrate capacity. The heat transfer performance is discussed and expected coefficients for typical duties given. In addition, prediction formulae are shown for both heat transfer and pressure drop.

NOTATION

A area of single pass of plate heat exchanger
B Nusselt number constant for plate heat exchanger
B_1 constant dependent on spacing—spiral heat exchanger
B_2 constant dependent on stud density—spiral heat exchanger

B_3 pressure drop constant—spiral heat exchanger
C friction factor constant—plate heat exchanger
D tube diameter
D_e volumetric hydraulic mean diameter—plate heat exchanger
D_v volumetric hydraulic mean diameter—finned-tube heat exchanger
f friction factor
g_n acceleration due to gravity
H_1 width of spiral heat exchanger
H_2 fin height—finned-tube heat exchanger
L length of flow passage
L_1 passage spacing—spiral heat exchanger
$\dot{m}$ mass velocity, i.e. mass flow rate per unit across sectional flow area
$\dot{M}$ mass flow rate
u fluid velocity
U overall heat transfer coefficient
$\dot{W}$ power
x gap between fins—finned-tube heat exchanger
y transverse pitch—finned-tube heat exchanger, i.e. distance between tubes in a transverse row
z longitudinal pitch—finned-tube heat exchanger, i.e. distance between adjacent tubes in two transverse rows
α film heat transfer coefficient
η dynamic viscosity
η_b bulk viscosity
η_w wall viscosity
λ thermal conductivity of condensate
ν kinematic viscosity
ρ fluid density
Δp pressure drop
ΔT temperature difference
Δh_v latent heat of condensation
Nu Nusselt number
Pr Prandtl number
Re Reynolds number

1. SCOPE

The term 'compact heat exchangers' is open to a number of different interpretations. Some studies[1] confine it broadly to gas-flow applications, where the need for an extended heat transfer surface on the gas side results in a high surface/volume ratio, and this particular parameter has been used as a means of differentiating between compact and non-compact units. Suggestions in the literature[2,3] for a dividing line vary between 330 m^2/m^3 and 700 m^2/m^3, whereas a comprehensive analysis of 89 different types of gas-flow heat exchanger[1] gives surface volume ratios ranging from 2700 to 150 m^2/m^3.

It is therefore difficult to make a firm distinction on this basis, and this chapter adopts a much broader interpretation which covers any form of construction of a direct-type heat exchanger, liquid–liquid, liquid–gas or gas–gas whose surface/volume ratio is significantly greater than that of an orthodox shell-and-tube unit, some of these nevertheless having ratios which are lower than the figures given above.

The types of heat exchanger covered in this chapter generally derive their compactness from one of two factors:

(a) The need for an extended heat transfer surface. These are mainly gas-flow heat exchangers of finned-tube types or plate-fin types for cryogenic applications.
(b) The use of plates instead of tubular heat transfer surfaces. This enables the surface/volume ratio to be drastically reduced depending on the pitch of the plates themselves, and is embodied in either gasketted plate, spiral or lamella types.

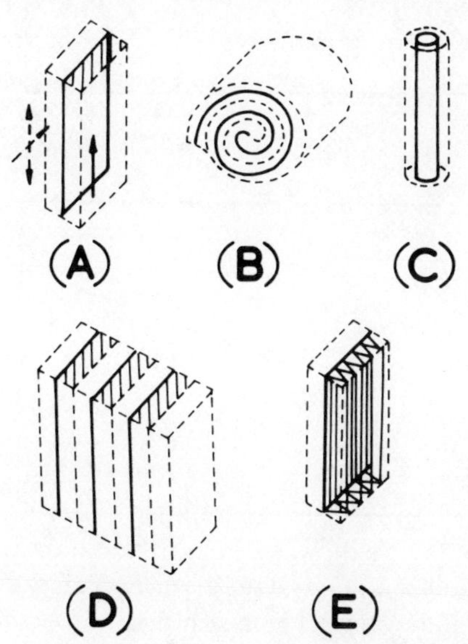

FIG. 5.1. The interrelation of compact heat exchanger configurations.

Outside these categories is the carbon block unit, which derives its compactness from the close spacing which can be obtained between the circular heat transfer passages in the block itself.

The inter-relation between different types of compact heat exchangers has been demonstrated by Patten[2] as shown in Fig. 5.1. The basic module is a flat plate (A) which separates the hot and cold fluids. When bent into a Swiss-roll configuration, it forms a spiral heat exchanger (B), while if bent around the other axis it forms a tube (C). A number of flat plates in parallel form an orthodox gasketted plate heat exchanger (D), and if alternate plates are welded together down their edges and enclosed in a cylinder, the lamella construction results. On the other hand, a plate module with a secondary corrugated plate attached to it demonstrates the construction of a plate-fin unit (E).

Compared with the many configurations adopted in compact heat exchangers, the tube has a simple geometry which enables internal film

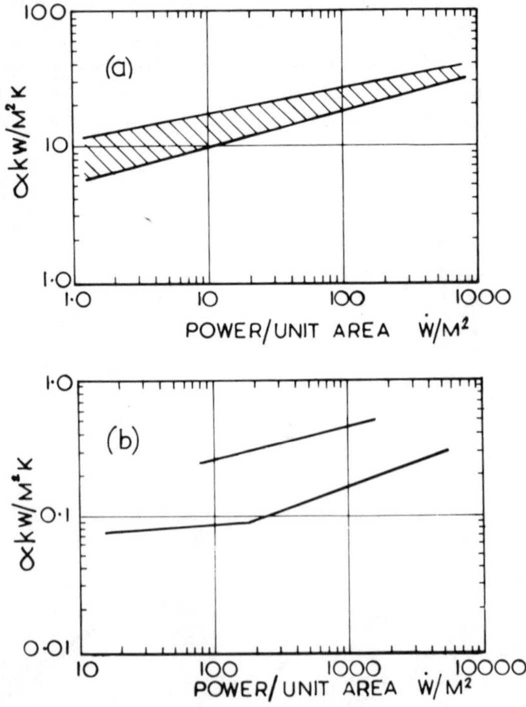

FIG. 5.2. Heat transfer coefficient versus power consumed for two typical compact heat exchangers.

coefficients to be predicted in terms of length and diameter according to well established relationships. Even so, coefficients on the shell side are much less predictable because of their dependence on tube arrangement, baffle type and spacing, etc. Certain types of compact unit also lend themselves to analytical treatment, but others have such complex geometry that this is not possible, and the performance characteristics must be established experimentally. However, a method of comparing different compact units of the same basic type is to plot film coefficient α as a function of the mechanical power $\dot{W}$ per unit surface area necessary to overcome fluid friction, those units with high coefficients being described as having a high performance or a high *surface goodness factor*. As an example of two different types, Fig. 5.2 shows typical $\alpha/\dot{W}$ plots for:

(a) Corrugated plate heat exchangers on aqueous liquid/liquid duties.[4]
(b) Different forms of extended surface plate-fin heat exchangers based on the fluid properties of air at 10^5 Pa and 260 °C.[1]

2. PLATE HEAT EXCHANGERS

2.1. Construction

2.1.1. Principle of Operation

The plate heat exchanger, first introduced commercially in 1924, developed from the requirement for a means of heat transfer for the process industries which would give complete accessibility to both liquid surfaces. This led to the concept of a number of rectangular plates clamped together in a frame, each sealed by a peripheral gasket and carrying four gasketted corner ports. Entry and exit connections for the two liquid media are made through the frame to the two pairs of top and bottom ports, and the gasket design is arranged so that the two liquids flow countercurrently through alternate passages between the plates, as shown in Fig. 5.3.

Two of the ports conduct one liquid in and out of one heat transfer passage, and the other two transfer the other liquid into the adjacent passage. The space between the transfer port of one liquid and the heat transfer passage of the other is vented to atmosphere at V so that it is impossible for direct leakage to take place across the gasket from one liquid to the other.

The plates themselves are pressed in light gauge metal and are corrugated in some way with the object of increasing their mechanical strength, and this results in the added advantages of improving heat transfer by inducing turbulence and of maintaining the correct flow space between the plates.

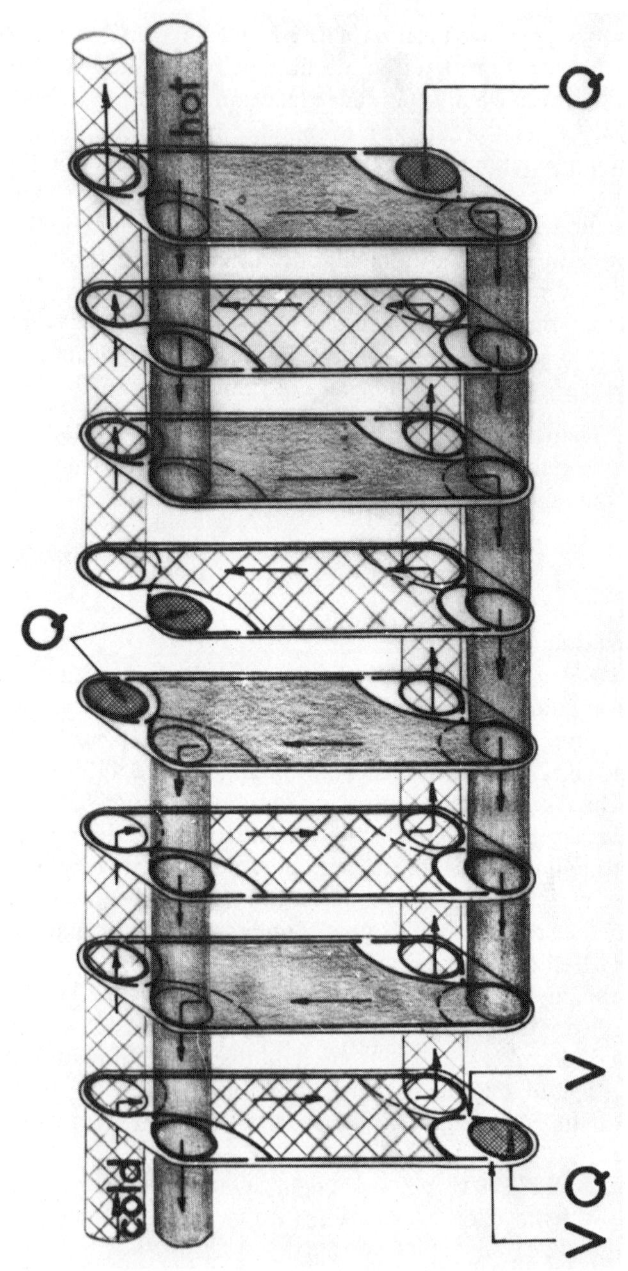

FIG. 5.3. A plate heat exchanger in 2-pass countercurrent flow.

Any number and size of pass can be achieved within the capacity of the frame, and Fig. 5.3 shows an example of an arrangement of two passes of two passages against two passes of two passages. The actual pass arrangement is obtained by terminating the end of each pass by solid rather than open ports at Q, and the system shown will have seven active thermal plates, with additional inactive plates at the ends to enclose the outer flow passages, so that in general:

No. of active thermal plates = (total no. of flow passages) – 1

This system also permits differing numbers and size of pass for each liquid, e.g. by a suitable choice of closed ports, Fig. 5.3 could be modified to two passes of two/one pass of four in order to reduce the pressure drop for the second liquid, but this would result in concurrent flow in one of the smaller passes.

By means of special connector plates, service fluids can be introduced at different points along the plate pack, allowing multiple duties, such as regeneration, holding, heating and cooling to be incorporated in a single frame.

2.1.2. Corrugation Design

A wide variety of corrugation patterns are used, the most common being:

(a) Transverse intermating corrugations pressed deeper than the plate spacing, which nest with those of adjacent plates, so giving a constant change of velocity and flow gap of a two-dimensional nature. These corrugations are surmounted with pips or dimples to maintain the flow gap.

(b) Corrugations pressed to the depth of the gasket, in the form of a chevron or herring-bone pattern, the angle of obliquity of the chevron being reversed on adjacent plates, so that when the plates are clamped together, interplate support is given where the corrugations cross and contact one another. The resulting flow pattern is three-dimensional, and this type is most commonly used because:

   (i) The large number of contact points enable higher pressures to be handled and lighter metal gauges to be used.

   (ii) The shallower depth facilitates pressing in the less ductile materials.

   (iii) A high degree of turbulence is produced, giving high heat transfer coefficients at low flow rates.

2.1.3. Frame

The frame takes the form of a press with the plates clamped between a fixed and a moving member made of carbon steel. The plates hang vertically on a top carrying-bar and are guided at their lower end by a bottom guide bar, both being supported between the fixed member and an end support. The clamping force is provided either by tie bars between the fixed and moving members or by one or two compression screws between the moving member and the end support, the former system being the simpler in construction because fewer stressed members are involved, but the latter facilitating ease of opening and closing of the frame. Frames used for hygienic duties are frequently clad in stainless steel, and means of power tightening are available.

2.2. Performance

2.2.1. Hydraulic

The main characteristic of the plate heat exchanger stems from the turbulence induced by the plate corrugations, which results in a low critical Reynolds number compared with the tubular figure of 2300. Plates may be designated *hard* or *soft* depending on whether they are designed for high or low turbulence, and the critical Re value may vary from 10 to 800 over this range.

In laminar flow

$$f = \frac{C}{\mathrm{Re}}$$

where C varies from 35 to 60 compared with 16 for tubes and 24 for flat ducts.[5] In turbulent flow, a typical relationship is

$$f = 2{\cdot}5\left(\frac{\dot{m}D_e}{\eta}\right)^{-0{\cdot}3}$$

both expressions applying to the equation

$$\Delta p = \frac{2f\dot{m}^2 L}{\rho D_e}$$

where D_e, the volumetric mean diameter is given by

$$\frac{4 \times \text{volume between plates}}{\text{wetted surface area}}$$

2.2.2. *Thermal*

Typical equations for the laminar[5] and turbulent[6] regimes respectively are:

$$\mathrm{Nu} = B\left(\mathrm{Re}\,\mathrm{Pr}\frac{D_{\mathrm{e}}}{L}\right)^{0\cdot33}\left(\frac{\eta b}{\eta w}\right)^{0\cdot14}$$

where B varies from 1·86 to 4·50 depending on plate geometry, and

$$\mathrm{Nu} = 0\cdot28\,\mathrm{Re}^{0\cdot65}\,\mathrm{Pr}^{0\cdot4}\left(\frac{\eta b}{\eta w}\right)^{0\cdot14}$$

2.2.3. *Overall Performance*

Plates normally operate over pressure drops up to 1–1·3 × 10^5 Pa, at which overall htc's of the order of 6500 W/m^2 K on water can be obtained, and performance can usefully be rated in terms of NTU's (number of transfer units).

The NTU value of a single pass of plates[7] is given by

$$\frac{2UA}{\dot{M}c_{\mathrm{p}}}$$

and on water, these NTU values vary typically between a maximum of 4 for 'hard' plates and 0·7 for 'soft' ones.

For any particular thermal duty, the NTU's required are given by

$$\frac{\text{Temperature rise}}{\text{LMTD}}$$

So that a thermal duty of 12 NTU's would require at least three passes of the 'hard' plate quoted above, the number of passes being adjusted to comply as far as possible with both thermal and pressure drop requirements. An alternative method of achieving this is by 'thermal mixing', a system which was introduced in the early 1970s and is based on combining packs of 'hard' and 'soft' plates with chevron corrugations which are mechanically compatible but which have high and low NTU characteristics. The relative proportions of the two packs are then sized so that the required thermal and pressure drop specification is accurately complied with.[8]

Total area and pass arrangements for PHE systems are calculated by normal heat transfer methods, but allowance must be made for the following correction factors:

(a) Concurrency resulting from unequal passes together with other factors which also depend on the pass arrangement.[9]

(b) Distribution effects due to pressure loss along the feed ports.[10]

The overall characteristics of the PHE mean that it is primarily suited for liquid–liquid duties, but the combination of high NTU values with general overall countercurrency and geometrically similar flow passages for each fluid gives it the ability to achieve very close end approach temperatures. For recuperative duties, PHE's are particularly suitable where efficiencies of over 92 % can be obtained. Vapour condensing duties can also be dealt with, but the finite size of the feed port limits the amount of vapour which can be handled.

2.2.4. Operating Conditions

Trends in design have mostly centred on larger plates and higher operating pressures in a wider range of materials, to meet the demands of industries such as the chemical, nuclear power and offshore oil industries.

Plates are now available in stainless steel, titanium and most ductile corrosion-resistant alloys. However, the use of rubber as a gasketting material limits both the maximum operating temperature and chemical resistance of the units, and in conjunction with the strength of the frame, also restricts the operating pressure. Gaskets of compressed asbestos fibre are available for some plates and allow increased temperatures and give greater resistance to chemical solvents.

Typical mechanical details are given below, though not all of these are applicable to the same machine:

Operating pressure	up to $2{\cdot}0/2{\cdot}75 \times 10^6$ Pa
Operating temperature	up to 170 °C for rubber gaskets
Plate heat transfer area	up to $2{\cdot}6\,\mathrm{m}^2$
Maximum heat transfer area for a complete machine	$1560\,\mathrm{m}^2$
Maximum port size	0·4 m diameter
Corresponding maximum flow rate	$0{\cdot}7\,\mathrm{m}^3/\mathrm{s}$
Plate thickness	0·6–1 mm
Plate flow gap	1·5–5 mm

3. SPIRAL HEAT EXCHANGERS

3.1. Construction

There are a number of types of welded plate heat exchangers which seek to remove some of the limitations resulting from the use of rubber gaskets in orthodox plate units, but a completely all-welded construction virtually eliminates any possibility of accessibility for internal inspection. The spiral design however avoids the use of rubber but at the same time allows some degree of disassembly.

It relies for its operation on two long metal strips which are wound in a spiral around a common axis, thus forming two parallel channels through which the two fluid media pass. The spacing between the two spirals is maintained by welded studs, and the whole assembly is housed in a cylindrical shell which is enclosed by top and bottom circular covers, either flat or conical.

There are three alternative forms of construction shown in Fig. 5.4, and they depend on the type of duty to be handled.

(a) Liquid–liquid duties (Fig. 5.4(a)). Here the spirals are welded together along their top and bottom edges alternately, so that when the unit is enclosed by flat covers (C) carrying an asbestos-based gasketting material (M) which seals along the edge welds, two enclosed parallel passages are formed, each passage being fitted with central and peripheral connections (A and B) resulting in true countercurrent flow of the two fluid media. Removal of the covers gives accessibility to both liquid passages and this type is also sometimes used for gas or vapour–liquid duties, provided pressure drops are not excessive.

(b) Low pressure condensing duties (Fig. 5.4(b)). In this type, the vapour flow (V) is transverse to the liquid (W) in the spiral, this being achieved by enclosing the liquid passage by welding both the top and bottom edges of the spirals and fitting conical covers (C) which allow the vapour to pass cross-flow to the liquid, thus minimising the pressure drop and facilitating the removal of condensate (L) and non-condensables (S). With this design, the liquid passage is completely enclosed and only the vapour passage is accessible.

(c) Condensing with sub-cooled condensate (Fig. 5.4(c)). This is used for condensing vapour–gas mixtures where it is necessary to cool both the gas (G) and the condensate (L). The top cover has a central cone which gives good distribution of vapour over the spiral, while the flat periphery is designed to seal off the top of the outer turns, so that the gas (G) and condensed vapour (L) leave countercurrent to the cooling medium, appropriate exit connections being fitted to the shell.

These units can be made from any weldable material of adequate strength, and two methods of enclosing the spiral passages are used. Originally a bar

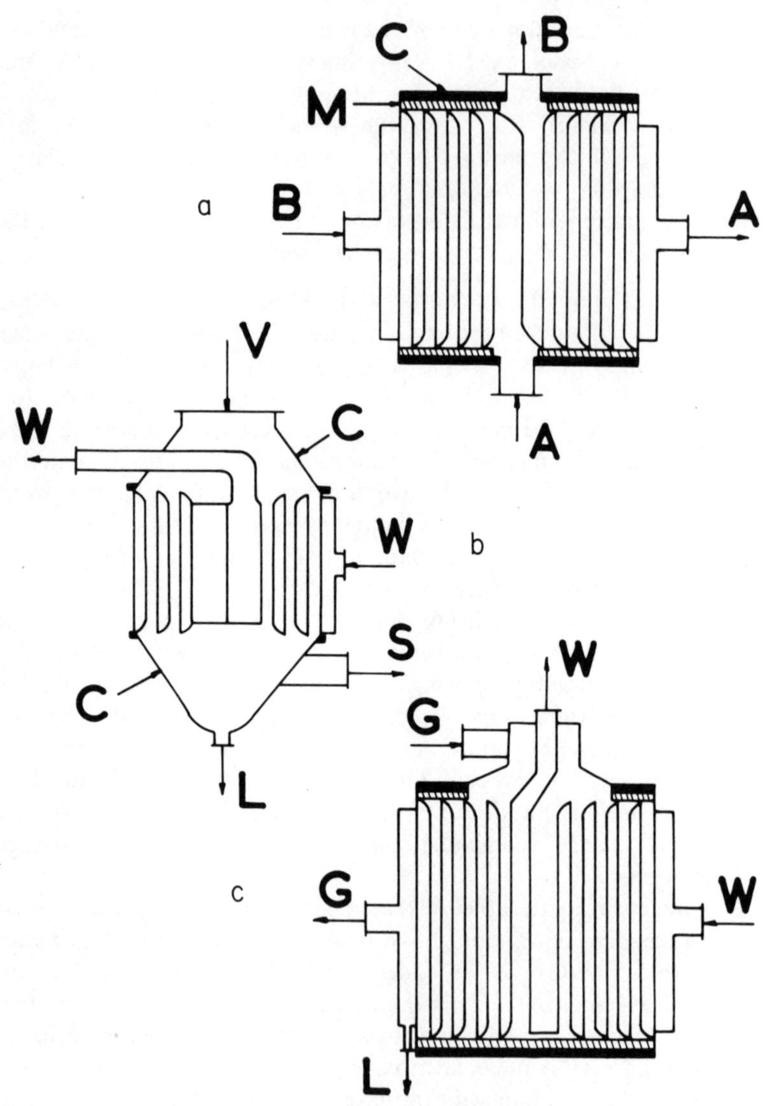

FIG. 5.4. The spiral heat exchanger—three typical configurations.

was welded between the two spirals, but it is now more usual to bend over the edge of one spiral so that it contacts the edge of the other, and then seal this contact by edge welding. It will be noted that the gasketted cover in Fig. 5.4(a) prevents by-passing of the spirals by one of the liquids, but that any mixing between them is prevented by a welded joint.

3.2. Operating Conditions

The spiral shares with the orthodox plate heat exchanger the advantage of true countercurrent flow, but its construction and the avoidance of rubber gaskets means that it will withstand higher temperatures. Cross-flow permits it to handle larger quantities of vapour and the fact that these units are custom-built means that both flow gap and width can be varied to suit any particular duty, unlike pressed plates where these dimensions are fixed. However, it lacks the flexibility of the plate heat exchanger, whereby any number of plates can easily be arranged to give a wide variation in size and number of passes or alternatively to incorporate thermal mixing, at the same time giving complete accessibility to the heating surfaces. In the case of the spiral, operation is confined to single pass, and the maximum NTU value obtainable is determined by the length of the spiral passage, which has a maximum of about 140 m, allowing NTU values of about 3–4 to be achieved on aqueous liquids. Technical Data are: maximum surface area 200 m^2, maximum pressure 15×10^5 Pa, maximum temperature 400 °C and passage spacing of 5–25 mm.

3.3. Performance

Work by Sander[11] has been reported which gives the following equation for heat transfer in turbulent and transition flow:

$$\mathrm{Nu} = \mathrm{Pr}^{0\cdot 25}\left(\frac{\eta b}{\eta w}\right)^{0\cdot 17} 0\cdot 0315\,\mathrm{Re}^{0\cdot 8} - 6\cdot 65 + 10^{-7}\left(\frac{L}{L_1}\right)^{1\cdot 8}$$

Pressure drop in turbulent flow is given by

$$\Delta p = \frac{Lu^2\rho}{B_3}\left[\frac{B_1}{\mathrm{Re}^{0\cdot 33}} + B_2 + \frac{16\cdot 4}{L}\right]$$

For condensing under cross flow conditions, the film coefficient is given by

$$\alpha = 4\cdot 2\left(\frac{\lambda^3 \rho^2\,\Delta h_v}{\nu H_1\,\Delta T}\right)^{0\cdot 25}$$

For fresh water duties, overall heat transfer coefficients range from 1200 to

2500 W/m^2 K and for condensing of steam from 1300 to 1800 W/m^2 K. At the other end of the scale, oil cooling gives figures of 170–520 W/m^2 K, and gas–gas and gas–liquid duties are covered by 20–70 W/m^2 K.

4. LAMELLA HEAT EXCHANGER

4.1. Construction

The lamella is another form of welded heat exchanger which seeks to give some accessibility by combining the construction of a plate unit with that of a shell-and-tube exchanger.

In this design, tubes are replaced by pairs of flat parallel plates which are edge welded to provide long narrow channels, and banks of these elements of varying width are packed together in an outer shell so that the flow area on the shell side is a minimum and similar in magnitude to that of the inside of the bank of elements. This means that the velocities of the two liquid media are comparable, as shown by the cross-section illustrated in Fig. 5.5. One end of the element pack is fixed and the other is floating to allow for

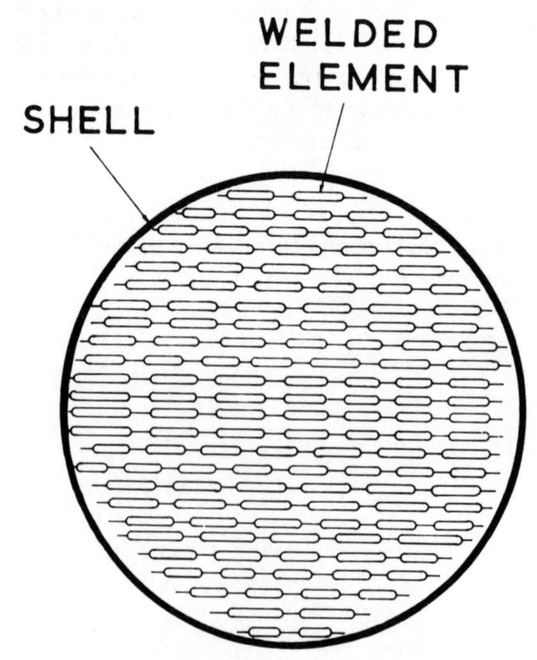

FIG. 5.5. A cross-section of the lamella heat exchanger.

thermal expansion, with connections fitted at either end of the shell as in the normal shell-and-tube design, a construction which allows the bank of elements to be withdrawn, so making the outside surface accessible.

4.2. Operating Conditions

The two process liquids flow countercurrently, so that the disadvantages of the shell side in the orthodox tubular heat exchanger are avoided, while the narrow flow gap results in a small hydraulic mean diameter, thus enabling relatively high heat transfer coefficients to be obtained. Because of the accessibility of the outside of the elements, this side is normally used for any fouling conditions, and also for boiling and condensing, where some local enlargement around the connections is sometimes introduced to minimise pressure drops.

Technical data are: maximum surface area 750 m^2, maximum pressure $3{\cdot}6 \times 10^6$ Pa, maximum working temperature 500 °C, and heat transfer coefficients (water–water) 1700–3400 $W/m^2 K$.

4.3. Performance

Film coefficients under turbulent flow follow a Dittus–Boelter relationship of the form

$$Nu \propto Re^{0{\cdot}8}\,Pr^{0{\cdot}3}$$

while pressure drop can be derived from normal parallel duct theory.

5. FINNED-TUBE HEAT EXCHANGERS

5.1. Construction

Heat exchangers described in previous sections have equal heat transfer areas for both fluid media. For duties such as the air cooling of industrial liquids, it is necessary to have both a forced draught and an extended surface on the air side to compensate for the low film coefficient.

Such heat exchangers are therefore based on the use of transversely finned-tubes, in which the fin normally takes the form of a helix, a section of which is shown in Fig. 5.6(a), and is formed by a variety of methods:

Mechanically Secured or Welded Fins

(a) For light duties and temperatures up to 150 °C, the fin has either a right-angled flange or is crimped at its base in order to provide a

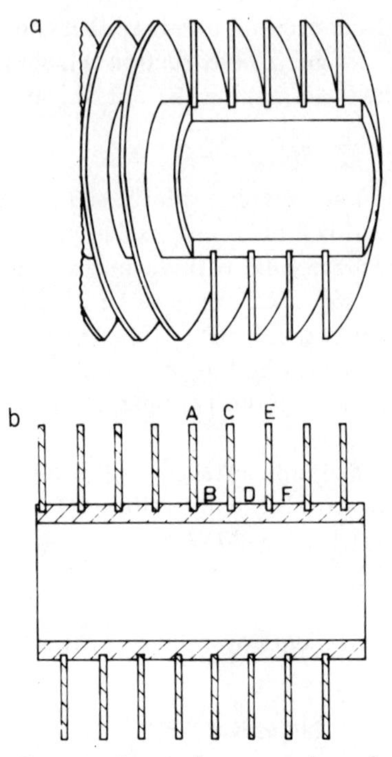

FIG. 5.6. Construction and cross-section of a finned tube.

firm contact with the outside of the tube when the fin is tension-wound into position, while for more severe conditions the fin is wound on to the tube and seated securely with knurling rolls which raise the tube metal on either side of the fin.

(b) For temperatures up to 400 °C, a helical groove is ploughed into the outside of the tube and a pretensioned fin tightly wound into it, after which the displaced metal is forced tightly against the base of the fin to give a firm seal. Temperatures above this figure require the fins to be welded to the base tube.

(c) Mechanical attachment is sometimes supplemented by soldering, galvanising or tinning in order to improve thermal contact, and all the above types obviously permit different combinations of materials for fin and tube, so that fins made of high conductivity metals such as aluminium, can be wound on to mechanically strong tubes of carbon or alloy steel, although problems of differential expansion must be considered. Where fin and tube metals are similar, flange-based fins can be resistance-welded to the tube, and one variant uses serrated fins which increase turbulence and improve heat transfer.

Integrally Finned Tubes
Fins are formed from the tube material itself by rolling, and are produced in two types:

(a) Low fin, in which the O.D. of the fin is no greater than that of the tube, thus facilitating assembly in an orthodox shell-and-tube construction.
(b) High fin, in which the O.D. of the fin is not restricted, and the maximum degree of extended surface is obtained. As an example, extruded aluminium fins are suitable for severe duties up to 260 °C. High fins also include certain bi-metallic combinations which are produced by metallurgically bonding the outer extended surface to the inner tube.

For air cooling duties, the tubes are mounted in headers which can take the form of a cast or welded box, a welded manifold or a box fitted with a bolted coverplate for accessibility. The whole assembly is then supported in a rigid frame which also carries the fan, which may be of the forced or induced draught type, depending on mechanical considerations.[12]

5.2. Design and Performance

The air- or gas-side performance of finned tubes has been studied by considering the two extreme conditions, heat transfer on the outside of an unfinned tube and heat transfer between flat parallel plates.[13] This has led to the following recommended correlations for fins and tubes of identical materials with a perfect thermal bond between them:

Plain fins

$$\mathrm{Nu} = 0{\cdot}1378\,\mathrm{Re}^{0\cdot718}\,\mathrm{Pr}^{1/3}\left(\frac{x}{H_2}\right)^{0\cdot296}$$

Corrugated fins

$$\mathrm{Nu} = 0{\cdot}056\,\mathrm{Re}^{0\cdot8}\,\mathrm{Pr}^{1/3}\left(\frac{D}{H_2}\right)^{0\cdot427}$$

The relation between pressure drop and friction factor is given by Gunter and Shaw[14] as:

$$\Delta p = \frac{f\dot{m}^2 L}{2D_v}\left(\frac{\eta b}{\eta w}\right)^{0\cdot14}\left(\frac{D}{y}\right)^{0\cdot4}\left(\frac{z}{y}\right)^{0\cdot6}$$

For laminar flow

$$f = 180\,\mathrm{Re}^{-1}$$

For turbulent flow

$$f = 1{\cdot}92\,\mathrm{Re}^{-0\cdot145}$$

where Re is based on the volumetric hydraulic mean diameter:

$$D_v = \frac{4 \times \text{nett free volume}}{\text{friction area}}$$

the friction area being defined by profile ABCDEF in Fig. 5.6(b).

However, in designing the complete heat exchanger it is necessary to consider the frontal area and depth of the complete unit so that the air velocity over the fins will comply with both the heat transfer and pressure drop requirements, while fin efficiency must also be taken into account.

Fin efficiency is the conventional expression which allows for the fact that the temperature of a fin, due to heat transfer through it, is different at the tip compared to the base. Thus, the expected gain in heat transfer due to increased area is partly lost due to a reduction in temperature difference. Fin efficiency is therefore defined as the average temperature difference over the extended surface compared with that over the basic surface. For many configurations the efficiency can be calculated from geometric considerations as shown by Gardner.[15] Thus a design calculation should take into account fin efficiency in the normal heat flux equation.

Tube length is an important criterion; in principle the longer the tube the better, within the constraints of tube-side pressure drop, which together with tube-side heat transfer coefficient is calculable by normal tube correlations. A small duty would not be best achieved in a long narrow exchanger except as part of a multi-duty unit where the fan arrangement allows efficient use of power. Choice of induced or forced draught exchangers depends on layout rather than heat transfer as the good distribution of an induced draught system is compensated for by extra turbulence in a forced draught system. A good 'rule of thumb' sizing calculation for air-cooled heat exchangers is given in reference 16.

Conventional design calls for a relatively small number of tubes in the air flow direction (four to six rows in triangular pitch) the air flow space shape being arranged to provide an acceptable tube-side pressure drop (dependent on length and number of tubes in parallel) and a shape which will accommodate one or a number of fans. Maximum tube length for mechanical and manufacturing viability is about 12 m.

6. PLATE-FIN HEAT EXCHANGERS

6.1. Development and Applications

The use of aluminium plate-and-fin heat exchangers has been established for nearly thirty years. Their compactness, with the associated low weight, low heat capacity, good ductility and increasing strength of the alloys involved at low temperatures, make them particularly suitable for cryogenic service.[17,18]

The plate and fin form of construction and the aluminium dip-brazing techniques were first combined for aircraft applications in which a high performance/weight ratio is essential. Extension of the technology and the manufacturing facilities has enabled these characteristics to be maintained for the much larger assemblies required for industrial plant, and designs can now be provided for any combination of gas, liquid and two-phase systems.

These heat exchangers have been established for use in air separation and helium liquefaction plants since the early 1950s. Ethylene plants date from 1958, and the early 1960s saw an expansion into 'nitrogen wash' ammonia plants, hydrogen purification plants and natural gas plants, while many other uses have been found and are being employed in the chemical industry. Apart from pure heat exchange, purification by deposition of solids and fractionation are both possible in this form of construction.

6.2. Construction and Operating Pressure

The basic element, or flow channel (Fig. 5.7) consists of two flat plates separated by a thinner corrugated sheet and two edge sealing bars. A number of such elements are stacked up to provide either a cross-flow or counter flow arrangement (Fig. 5.8). The corrugated sheet and bars are attached to the flat plates by dip-brazing, the brazed joints providing both the thermal bonds for heat transfer and the mechanical bonds necessary to enable the flat plates to contain the stream pressure.

Four principal types of corrugation are available:

(a) Plain straight-through passages.
(b) Corrugations, as above, perforated by small holes.
(c) Continuously waved herring-bone passages.
(d) Serrated or multi-entry corrugations of a staggered form.

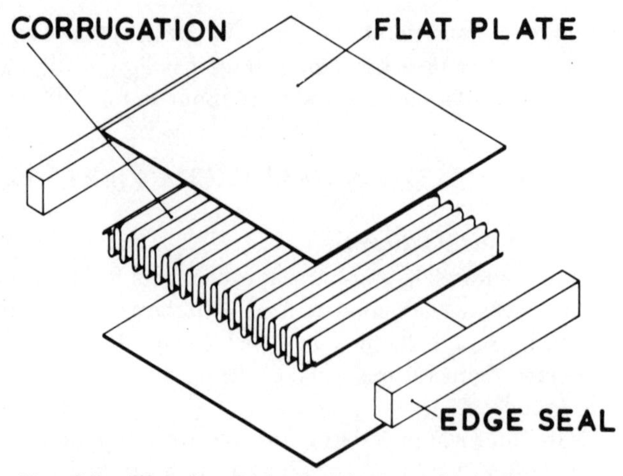

FIG. 5.7. The plate-fin heat exchanger—basic element.

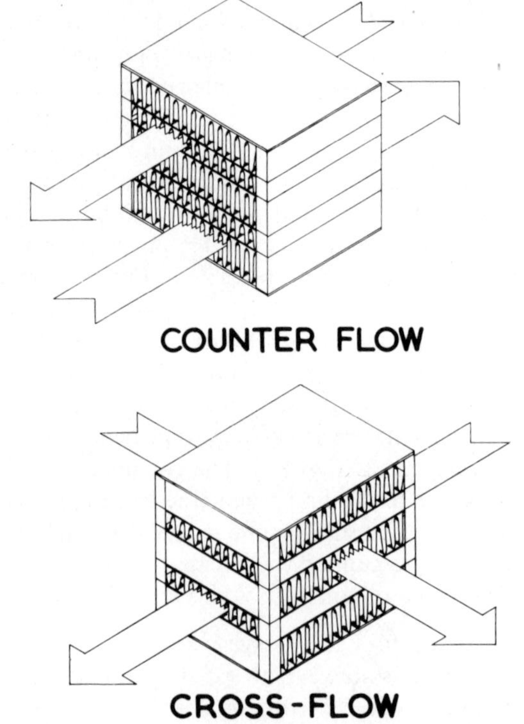

FIG. 5.8. The plate-fin heat exchanger—flow arrangements.

These units normally have a surface/volume ratio (m^2/m^3) in the order of 1230, with the following dimensional variants: Corrugation height 3·8–11·8 mm, corrugation thickness 0·2–0·6 mm, and fin density 230–700 fins/m.

Maximum Size

Counter flow units can be brazed up to 1·2 m wide, 1·23 m stacked height and 6·2 m long, giving a total area of about 10 600 m^2.

Maximum Pressure

The above unit will withstand a steady pressure of 25×10^5 Pa, although under reversing conditions on air separation plants, 10^6 cycles over approximately 20 years life reduces this to 11×10^5 Pa. Attention to the major determinants controlling operating pressure, such as stresses within the corrugations and their brazed joints and the junction of the header tanks to the matrix, enables design pressures of up to 80×10^5 Pa to be achieved depending on the overall cross-section.

Mechanical design is generally based on the rules of the A.S.M.E., but other specifications can be accommodated where appropriate.

6.3. Performance

The serrated and herring-bone corrugations give substantially increased heat transfer coefficients as compared to straight tubes or plain corrugations, the serrated type giving an increase of about four times the plain tube figure. Obviously the friction factors are also increased, such that a greater cross-section is required to maintain the same total pressure drop for a given thermal duty. Nevertheless there is a significant reduction in surface area and consequently in matrix volume.

Apart from this, the basic construction allows considerable versatility in design. Corrugation types may be chosen differently for various streams, as, for instance in liquid–gas heat exchangers where shallow and deep corrugations may be used for the liquid and gas streams respectively. Alternatively, two layers in the gas stream may be used for every single layer in the liquid stream, or different types of corrugation may be used in the two fluid media.

6.4. Multi-Stream Units

Another design variant enables several streams to be processed simultaneously. Each stream is allocated a suitable number of passages and the headering system at the ends of the matrix permits five separate streams to be accommodated with ease. This is of particular advantage in the field of cryogenic processes. Owing to the thermal bond linking plates and corrugations throughout the matrix, there is a continuity of metal temperature in each direction. Provided that each individual stream is evenly distributed across the whole matrix cross-section the separation plate temperature will also be reasonably constant over the whole cross-section. The local metal temperature controls the heat transfer rate for each separate stream. If several cold streams enter the heat exchanger at different temperatures the coldest streams will heat up more rapidly due to the greater temperature difference between that stream and the metal.

It is quite feasible under these circumstances to include a 'cold' stream in the heat exchanger which has an inlet temperature higher than that of the warm stream outlet. This return stream would be slightly cooled at inlet, before starting to warm up.

The basic requirement is that the weighted mean of all cold stream inlets must be below the required warm stream outlets. Where widely differing conditions are specified, often including two-phase operation, a step-by-step analysis may be required.

A further variety of multi-stream unit can be obtained by placing bars across the width of the block, thus enabling one stream to be taken out part way down the length of the matrix and permitting another stream to be inserted at the other side of the bars in the same passages. This enables several 'hot' streams at different temperature levels to be cooled by the same group of 'cold' streams in one matrix, with consequent savings in pipework and pressure drop.

7. GRAPHITE BLOCK HEAT EXCHANGERS

7.1. Advantages

The use of graphite for heat exchangers arises from the inert nature of its chemical properties together with its high thermal conductivity, which is approximately three times that of steel. Shell-and-tube units in this material, however, are very space consuming and an alternative concept of design is based on holes perforated in carbon blocks, where various forms of construction are available. Because of the nature of this construction, these have a low surface/volume ratio by conventional standards, of somewhere around 65–85 m^2/m^3 [19] but nevertheless they are more compact than the comparable shell-and-tube design and are mechanically stronger.

7.2. Impregnation

For most chemical plant, it is necessary to impregnate the graphite in order to render it impervious. The impregnant imparts no significant increase in the mechanical properties of the graphite,[20] but it does affect its resistance to corrosion and temperature. It is important that it[21]

(a) has chemical and thermal stability approaching that of graphite under the conditions of ultimate use.
(b) may be introduced into the graphite pores in such a way as to give fully repeatable results at a reasonable cost.
(c) will not set up stresses which would weaken the graphite structure during the impregnation/polymerisation process.

Of the three main impregnants, furanic and phenolic resins meet most normal requirements, but PTFE-based impregnants in the form of a colloidal solution will increase the temperature resistance to 220 °C or higher.

The impregnation stage is important as it determines the quality of the graphite and its suitability for any chemical process, and in one method[21] graphite is placed inside an autoclave which is then evacuated of air and adsorbed gases, after which the resin is admitted under vacuum. The autoclave is then pressurised for a while before being drained of resin. Thermal polymerisation follows whereby the autoclave is held at a high temperature for a prolonged period, and this process is repeated until satisfactory impregnation is obtained.

7.3. Mechanical Construction

In the design of this type of unit, it is important to consider the anisotropic nature of graphite, which means that the thermal conductivity is higher in one direction (along the planes) than in another (between the planes). Blocks must therefore be perforated so that the thermal conductivity is most effectively used.[22] Factors which must be taken into account are the high mechanical strength of graphite in compression and its low strength in tension. There are two principal forms of construction:[23]

(a) *Cubic/Rectangular Block*

The block is perforated with parallel holes between two pairs of opposite faces (one pair for each liquid) and is reinforced by end plates which are clamped together by tie bars at their corners as shown in Fig. 5.9.

Headers bolted on to the four faces of the block supply and remove the process fluids, and their design permits various pass arrangements to be

FIG. 5.9. Cubic heat exchanger. (By courtesy of Robert Jenkins Systems Ltd.)

obtained, with the object of approximating as far as possible to true countercurrent flow. Header materials must be compatible with process requirements, and graphite, steel or rubber linings can be used.

Standard units are available with heat transfer areas up to 15 m^2, and working pressures vary up to 5×10^5 Pa, depending on the design.

(b) *Modular Block*

In this design, illustrated in Fig. 5.10, a modular construction is used, in which the core is formed by a number of cylindrical drilled or perforated blocks compressed between two steel end plates by spring loaded tie bars and sealed with PTFE rings. Basically the process stream flows directly

FIG. 5.10. Polybloc (modular block) graphite heat exchanger. (By courtesy of Robert Jenkins Systems Ltd.)

from top to bottom and the service stream from bottom to top passing sideways through each block which therefore operates in cross flow, but the greater the number of blocks, the nearer is the approximation to overall true counterflow. The provision of a wide range of blocks with different heat transfer and flow areas gives a flexibility in design which enables the required duty to be achieved, while the unit can be easily dismantled for maintenance and the surface area modified for any change in duty.

Module areas up to about $10\,m^2$ enables units to be built with areas up to $250\,m^2$ and more in certain cases and, in general, design pressure of 6–8 $\times 10^5$ Pa. Higher pressures up to 20×10^5 Pa can be achieved, but for larger surface areas (above $150\,m^2$) conventional graphite shell-and-tube designs are usually more economic.

The short tortuous flow passages embodied in heat exchangers in this category help to induce turbulence, which in turn assists in minimising fouling and in the achievement of a good heat transfer performance.[24,25,26]

7.4. Performance

Graphite block heat exchangers are used for heating, cooling, evaporation and full and partial condensation, the fluids handled including inorganic acids, organic solvents and chlorinated organics, together with foods, drugs and other products where no metal contamination is permissible. Due to its excellent resistance to HCl in all states and the ability of graphite to conduct heat away quickly, the exchangers find wide use on hydrogen chloride absorption duties which are accompanied by evolution of heat.

Maximum passage diameters are approximately 20 mm, and figures published in the literature[23] give overall heat transfer coefficients of 1700–2300 $W/m^2\,K$ for the steam heating of aqueous liquids and of 850–1150 $W/m^2\,K$ for the cooling of 10 % hydrochloric acid, both at liquid velocities of 1·2 m/s.

For gas cooling (30 % hydrochloric acid, 7 % air with a trace of water vapour) at a velocity of 13·7 m/s coefficients of about 150 $W/m^2\,K$ are achieved while the evaporation of weak aqueous solutions and the condensation of water vapour at 100 °C give coefficients of 2000–2800 and 2000–2500 $W/m^2\,K$ respectively.

REFERENCES

1. KAYS, W. M. and LONDON, A. L. *Compact Heat Exchangers*, 2nd Edn., 1964, McGraw-Hill, New York.

2. PATTEN, T. N.E.L. Report No. 482, 1971, East Kilbride, Glasgow, National Engineering Laboratory, 1–6.
3. SHAH, R. K. and LONDON, A. L. Compact Heat Exchangers, Proc. 23rd Ann. Int. Gas Turbine Conf., 1978, London.
4. MARRIOTT, J. *Chem. Eng.*, 1971, **78**(8), 127–34.
5. PARROTT, D. Proc. 14th National Heat Transfer Conference, A.I.Ch.E. Reprint No. 5, 1973, A.I.Ch.E., A.S.M.E. Atlanta, Georgia.
6. COOPER, A. *Chem. Eng.*, 1975, **285,** 380–85.
7. CATTELL, G. S. Proc. Conf., Heat Transfer and the Design and Operation of Heat Exchangers, 1974, Johannesburg, S. Africa, CATT 1–CATT 28.
8. MARRIOTT, J. *Chem. Eng. Prog.*, 1977, **73**(2), 73–78.
9. FOOTE, M. R., N.E.L. Report No. 303, 1969, East Kilbride, Glasgow, National Engineering Laboratory.
10. WILKINSON, W. L. *Chem. Eng.*, 1975, **285,** 289–93.
11. SHINN, B. E. N.E.L. Report No. 482, 1971, East Kilbride, Glasgow, National Engineering Laboratory, 27–32.
12. GARDINER, J. E. *Proc. Eng.*, 1967, Nov./Dec., **6**.
13. SHERWIN, K. *Recent Developments in Compact High Duty Heat Exchangers*, 1972, I.Mech.E., London, 47–52.
14. GUNTER, A. Y. and SHAW, W. A. *Trans. A.S.M.E.*, 1945, **67,** 643–60.
15. GARDNER, K. A. *Trans. A.S.M.E.*, 1945, **67,** 621–32.
16. CROKE, R. J. *Practical Aspects of Heat Transfer*, 1978, A.I.Ch.E., New York, 168–69.
17. LENFESTEY, A. G. *Progr. Cryog.*, 1961, **3,** 24–47.
18. LENFESTEY, A. G. *Brit. Chem. Eng.*, 1960, **5,** 27–32.
19. BULL, J. L. *Recent Developments in Compact High Duty Heat Exchangers*, 1972, I.Mech.E., London, 67–76.
20. HILLIARD, A. *Chem. Proc. Eng., CPE-Heat Transfer Survey*, 1970, 59–62.
21. HILLIARD, A. Third Conference on Industrial Carbons and Graphite, 1970, Soc. Chem. Ind., London, 562–64.
22. HILLIARD, A. *Ind. Chem.*, 1963, **39,** 525–31.
23. HILLS, D. E. G. *Chem. Proc. Eng., CPE-Heat Transfer Survey*, 1971, 38–40.
24. HILLIARD, A. Proc. Eng. Heat Transfer Survey, 1974, 39–41.
25. WEYERMULLER, G. and LLOYD, F. R. *Chem. Proc. Chicago*, 1959, **22**(9), 44–45.
26. HILLIARD, A. *Chem. Proc. London*, 1965, **11,** 66–71.

Chapter 6

AIR COOLERS

C. D. R. NORTH
GEA Airexchangers Ltd, London, UK

SUMMARY

This chapter considers the design and application of air-cooled heat exchangers to process cooling requirements. The chapter is split into four sections covering major parameters affecting design and construction which need to be considered to determine the suitability of air-cooled heat exchangers for a particular application. The first section considers various important design temperatures and draws particular attention to freezing problems in low temperature areas. The second section considers the merits of the various configurations of air coolers with particular regard to the location of units. The third section considers air-side design; the need to allow adequate margins to achieve design airflow and the increasing importance of noise emission in determining fan and drive design. The final section briefly considers design and construction of air coolers and the merits of various header types, and methods of fabrication.

1. INTRODUCTION

This chapter considers the important parameters which govern the selection and thermal design of air-cooled heat exchangers. The air cooler is a major item of process equipment competing directly with water cooling on sites throughout the world. In general the capital cost of air coolers is higher, but the running cost is lower than the cost of providing treated water for shell and tube units. Where seawater is available the air cooler competes with titanium plate exchangers on indirect service in conjunction with shell and tube units, thus many of the large cooling services on oil rigs in the North Sea use direct air cooling.

FIG. 6.1. Typical air cooler installation.

A typical air cooler installation on a refinery piperack is shown in Fig. 6.1. As can be seen from Fig. 6.1 an air cooler is a combination of:

(a) Tube bundle.
(b) Plenum and structure.
(c) Fan and drive.

Mechanical design of an air cooler draws together the disciplines of:

(a) Pressure vessel design.
(b) Structural engineering.
(c) Rotating equipment.

The detailed methods of mechanical design are not discussed in this chapter. The engineer is referred to the relevant national and international design codes for such equipment.

This chapter is divided into four sections dealing with major parameters affecting the design of air-cooled heat exchangers:

(1) Design temperature.
(2) Exchanger configuration.
(3) Airside performance.
(4) Design and construction.

2. DESIGN TEMPERATURE

Several temperatures are important in the design and selection of air coolers.

2.1. Ambient Design Air Temperature

All air-cooled heat exchangers, except certain units with a cooling tower arrangement in series, are designed on the basis of dry bulb temperature. Seasonal variations as well as the variation from day to night are important and must be taken into account in the selection of the design ambient temperature by the process engineer.

The use of air coolers varies from essential services to incidental cooling of products to some safe temperature for storage.[1] Under-performance of units in critical service on a refinery could result in the need to flare to atmosphere undue amounts of toxic or corrosive fluids. Thus it is becoming common to utilise two design temperatures on a particular site, one for critical and the other for non-critical services. The latter is often taken as the temperature that is not exceeded for more than 5% of the three hottest months of the year. This information is generally available in standard meteorological tables. The resulting design temperature ranging from 20 °C in northern latitudes to 45 °C in the deserts of the Middle East can be 5–10 °C below the temperature necessary to guarantee operation under all ambient conditions. A further effect which must be considered in the selection of the design air temperature is the proximity of a bank of coolers to another major heat source. A large furnace can add 2–5 °C to local plant ambient temperatures. A similar problem arises with the bad location of the unit in the lee of a major structure. Rose, Wilson and Cowan[2] recommend certain sensible criteria for the location of air coolers.

Particular care should be taken by the design engineer when a unit is to be installed on an existing crowded site both as regards positioning too close to large structures and to take into account the prevailing wind and the elevation of the cooler relative to existing units. If sufficient separation cannot be achieved then the units should be wind skirted or sealed one to another to prevent hot air re-circulation.

For vacuum steam condensers associated with power generation, the peak load does not occur within the maximum summer ambient temperature and in this case design temperatures of up to 5 °C below standard are common with appropriate reduced performance guarantees in summer. Thus three design temperatures are possible:

(a) Low design temperature for certain power generation applications.

(b) Normal design air temperature.
(c) Elevated design temperature for critical service.

As discussed earlier, air coolers are sometimes used in conjunction with humidification systems in areas with low humidity and high temperature. Typical installations are high-pressure natural gas coolers in the Middle East where direct cooling is desirable. Two solutions are possible:

(a) Spraying condensate into air passing over the bundles.
(b) Using crossflow cooling sections prior to passing the wet air through drift eliminators and the cooler bundles.

Because of the high cost and limited availability of condensate, direct spray cooling of air is usually limited to emergency cooling and a special reservoir is kept for this purpose. With Solution (b) it is possible to use brackish water, but all the maintenance and 'digging out' associated with cooling towers is necessary. With this system the humidification section is only used during the summer period. During the winter either the air passes through the pre-cooling sections with no water flow or the sections are by-passed.

2.2. Minimum Ambient Design Temperature

The minimum ambient temperature affects both the selection of suitable materials for construction of the cooler and also the degree of freeze protection required to prevent plugging or damage to the tubes. Minimum temperatures as low as $-60\,°C$ are possible in Siberia, northern Canada and Alaska. The possibilities the design engineer has to consider to winterise this cooler are:

Process Control

In certain cases simply controlling the process outlet temperature with fan switching, two-speed fan drives, automatic louvres or autovariable pitch fans is sufficient.

Steam Coils and Louvres

In other cases the problem is limited to the start-up condition where low load on a cold unit could lead to freezing. A steam coil is a lightly finned or bare tube unit able to bring the cooler bundle to a safe working temperature with both process and fans off. Louvres (manual if no further process control is required) should always be fitted and kept closed to minimise losses due to natural draft.

According to the recommendation of API 661[3] this steam coil should be separate from the main cooler bundle, however, units integral with the bundle but independent on the process side appear to work well.

It is worth considering a protective coating on the core tubes of the steam coil as the unit will remain idle for most of its life. Suitable methods of protection include:

(i) Electrogalvanising.
(ii) Aluminising.

Both give a corrosion-resistant finish for a small capital outlay.

Co-current Flow

For some units, particularly viscous oil coolers, co-current design is appropriate. Here the hot fluid enters nearest the cold ambient wall, and the outlet where wall temperature is critical is in the warm air-stream. It is recommended that the tube wall temperature be maintained at least 10 °C above the pour point for freeze protection. Where no minimum flow is specified, 50 % turndown should be assumed. Where the outlet Reynolds number at turndown is likely to be transitional or laminar, it is important that the calculation be made by a point by point and not a mean coefficient method.

Unit By-pass

By-pass used either alone—to maximise cooler outlet temperature—or in conjunction with any of the above, may help to provide freeze protection.

Hot Air Re-circulation

As a principle this is a very elegant, albeit expensive way of achieving freeze protection. In practice there are many variations and no standard configuration has yet emerged. The principle is simple; if the air is too cold in winter, then some of the hot air is bled off the bundle through a system of ducts and mixed in a 'cabin' with the incoming air. To perform the calculation, the cooler must first be rated for the mass airflow to achieve the design duty at a safe temperature.

For process units containing water, a re-circulation temperature of about 10 °C is chosen. Only in the case of high pour point oils is there the need to re-circulate at or above summer design ambient temperature. Then the re-circulation system losses must be added to the normal bundle and plenum losses and the full airflow used in the following calculations (Fig. 6.2) where: x = % re-circulation, T_A = winter ambient °C, T_O = outlet air temperature °C, T_{rec} = required re-circulation temperature °C.

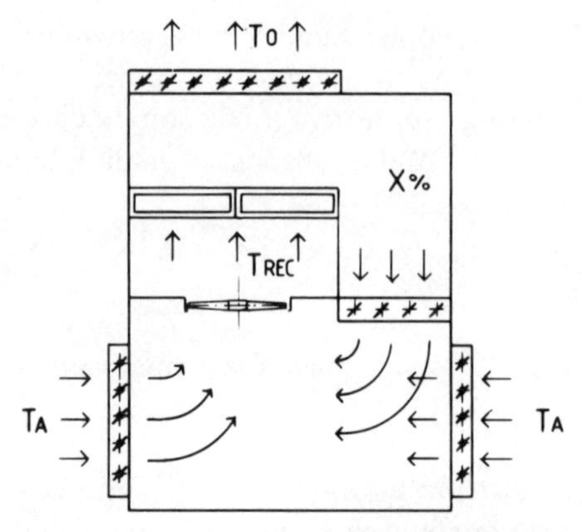

FIG. 6.2. Hot air re-circulation schematic.

A very simple equation can then be set up to calculate the % re-circulation required.

$$x = 100(T_{rec} - T_A)/(T_O - T_A)\,\%$$

From the % re-circulation, the mass flow of air to be re-circulated and hence the duct size can be calculated. Reference to the fan characteristics (Fig. 6.3) shows that for a constant blade angle reduced airflow corresponds to an increased static head. As the reduced volume to maintain performance has already been calculated, the driving head to power the re-

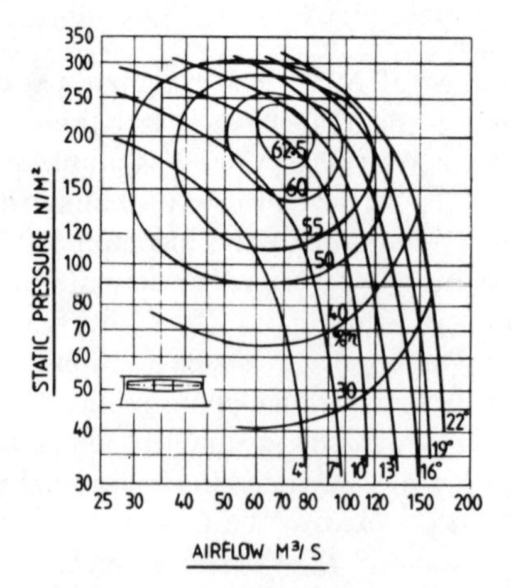

FIG. 6.3. Fan performance curve.

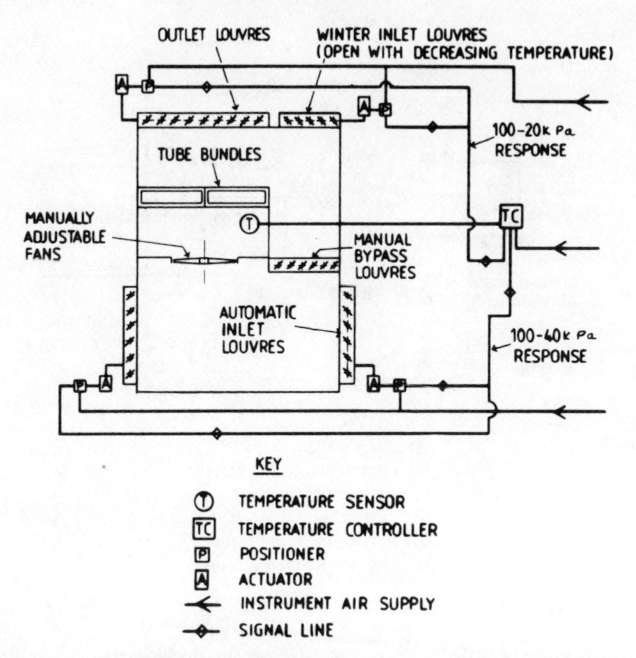

FIG. 6.4. Re-circulation system.

circulation system can be obtained directly from the constant blade angle line on the fan curve

$$H = \Delta P_{\text{rec}} - \Delta P_{\text{des}}$$

where H is the available head to power the re-circulation system. Depending on the selection of louvres and ducts, the pressure drop in a re-circulation system is likely to be in the range 3–6 duct velocity heads. Typical re-circulation systems are shown schematically in Figs. 6.4 and 6.5. Both systems feature solid cabin floors for ease of maintenance. In Fig. 6.5, top louvres and downcomer louvres operate in the opposite sense using pneumatic actuators. The bottom cabin outlet louvres are either set manually with the louvres nearest the prevailing wind closed in inclement weather or they are ganged against the opening downcomer louvres.

The temperature is usually measured by a long capillary tube within the plenum chamber after the airflows have been mixed by the fan. In Fig. 6.4 top inlet louvres are fitted in addition so that the louvres in the bottom cabin can be completely closed in winter. Often autovariable pitch fans are fitted

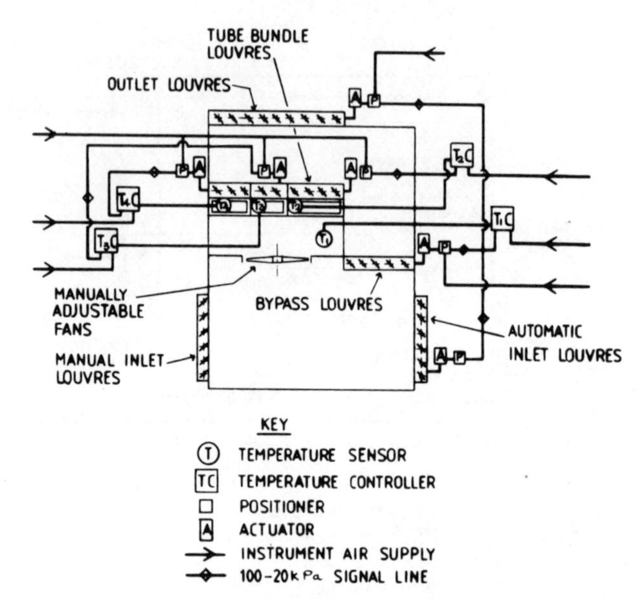

FIG. 6.5. Re-circulation system enclosing multiple duties.

to re-circulation systems. They should be used to control process outlet temperature to take up variations in load. They should not be interconnected with the control circuits associated with re-circulation temperature.

3. EXCHANGER CONFIGURATION

Often the choice of exchanger configuration is left to the thermal designer. However, in some situations the choice is critical to the proper operation of the plant and it is important that the project engineer understands the advantages and drawbacks of the various configurations.

3.1. Forced Draft–Horizontal Bundle

Advantages

(a) As the fans of a forced draft unit (Fig. 6.6) are in the cooler airstream below the bundle, the power consumption to pass a given mass of flow of air is lower.

(b) The fan drives are located in the cooler airflow below the unit, easing the problem of maintenance.

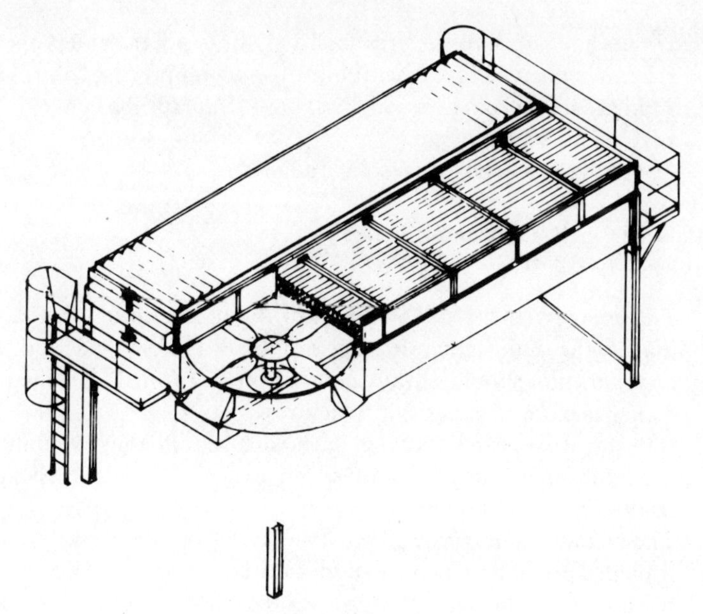

FIG. 6.6. Forced draft air cooler.

(c) The fans are unlikely to reach high temperatures unless close to a very high temperature bundle or in a re-circulation cabin, and thus construction materials are not critical.
(d) The erection procedure is simpler with the bundles located above the plenum chamber. Bundles can usually be removed for cleaning or repair without the need to disassemble structures.

Disadvantages

(a) Unless the unit is grade mounted, underslung walkways will be required for motor and fan access.
(b) The escape velocity of the air from the top of the bundle is low: typically 2·5–3·5 m/s. This makes the unit susceptible to crosswind effects inducing external re-circulation around the cooler. This problem is accentuated by proximity to tall structures or to other units not part of the same continuous bank. As discussed earlier, anti-re-circulation fences may have to be fitted at considerable expense.
(c) Good airflow distribution is more difficult to achieve than with induced draft exchangers.

(d) Unless hail screens or louvres are fitted, the top of the unit is open to the atmosphere. This presents a problem with process control as a sudden squall of rain or hail can dramatically change the performance of the unit.
(e) The bundle is exposed to solar radiation.

3.2. Induced Draft–Horizontal Bundle

Advantages (*Fig. 6.7*)

(a) The escape velocity of the air from the fan is up to 10 m/s, this makes the unit much less susceptible to crosswind and this configuration should always be used on a crowded site when the cooler is close to other tall structures.
(b) If the fan drives can be fitted in the hot air-stream, thus avoiding the need for underslung walkways and drives, this is the cheapest configuration.
(c) The plenum chamber whether of the 'hood' or 'flat decked' variety provides protection against sudden surges in performance caused by rain or hail storms; this improves the controllability of the unit.
(d) Cases have been reported on close approach units where solar

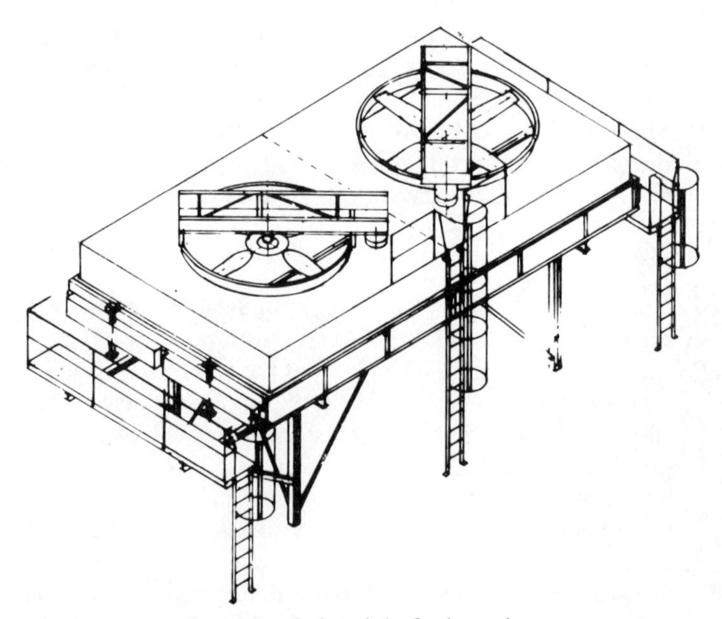

FIG. 6.7. Induced draft air cooler.

radiation on the bundles has added 5–10 % to the heat load. For this reason, the majority of units installed in the Middle East are induced draft.

(e) The plenum chamber and fan ring act as a chimney providing a higher heat rejection under fan failure conditions.

Disadvantages

(a) Except for very low process temperatures (below 70 °C) it is not appropriate to mount the gearbox or belt drive system in the hot air-stream.
(b) All autovariable fans, except those with remote actuation have low maximum operating temperatures and are not suitable for mounting in hot air-streams.
(c) All manual plastic fans and many manual aluminium fans with rubber, etc., in their attachments, have low temperature limits. The unit must be rated under maximum process temperature and total motor failure to ensure that the fan is suitable for the service.
(d) Because the fan operates in a warm air-stream the power consumption for a given thermal performance will be higher.
(e) Because the fan is designed to operate at a higher temperature, the margin between the motor power and the summer power will need to be larger to allow for cold starting in winter. A basic margin for the motor must consist of transmission losses plus the ratio of maximum summer operating to winter design temperature (K).
(f) Access to louvres is more difficult, as to be effective they need to be located between the tube bundle and the plenum chamber.
(g) The fitting of steam coils to an induced draft unit requires a skirt of at least 1 m to provide an equivalent to the plenum chamber associated with forced draft units.

3.3. 'A'-Frame Configuration

In this arrangement, cooler bundles are mounted on a triangular frame with forced draft fans below. The bundles are usually sloped between 45 and 60 ° from the horizontal (Fig. 6.8).

Advantages

(a) Minimises plot area.
(b) For vacuum steam condensers it is the basis of several patented 'non-freezing' systems.

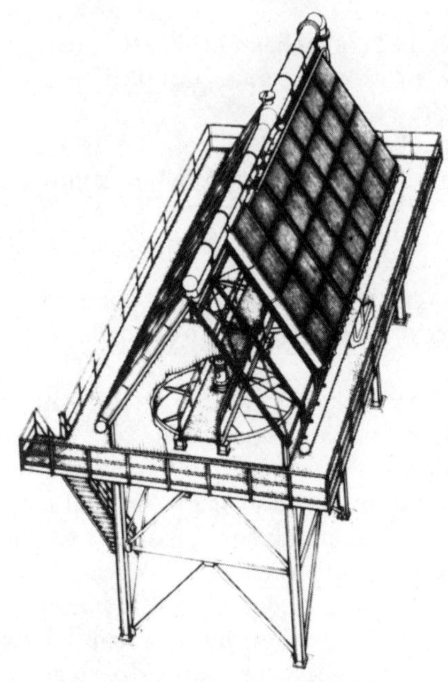

FIG. 6.8. 'A' frame exchanger.

Disadvantages

(a) More prone to hot air re-circulation.
(b) Prone to crosswind effects in bad weather conditions: it may require a wind/anti-re-circulation fence to prevent this.
(c) If the tubes are vertical, multi-pass bundles are not possible for condensing services; this can be mitigated by the bundles being in series.

3.4. Vertical Coolers

In this case, the bundles are at or near vertical with forced or induced draft fans (Fig. 6.9).

Advantages

Enables units to be mounted directly on columns, etc., e.g. column top reflux condensers.

Disadvantages

(a) Prone to crosswind effects.
(b) Multipass condensers are not practical.

FIG. 6.9. Column top condenser.

4. AIR-SIDE PERFORMANCE

Whilst bundle performance for a given airflow is a science, the design of plenum, fan and drive is still very much a 'Black Art' with arbitrary factors applied to cover various losses and maldistributions. In this section the necessary static head to ensure the passage of the required volume of air through the air cooler bundle is discussed. It is assumed that the bundle is properly sealed to the structure and that adequate seals are fitted to the bundle itself to prevent by-passing and wasting of airflow.

Most fan performance curves are drawn for the ideal situation at the entrance to a long duct, thus margins to allow for non-ideality must be incorporated before plotting on a standard fan curve.

4.1. Bundle Pressure Drop

There are two main sources of data for pressure drops through banks of fin tubes:

(a) Specific wind tunnel tests.
(b) General correlations based on a large spread of data.

For the high fin tubes most generally associated with air coolers the most-used correlation is that of Robinson and Briggs.[4]

This correlation is robust and extends beyond the most common range of airflow, i.e. 8000–12 000 Reynolds number.

$$DP_B = \frac{f_a G_m N_r}{\rho_a}$$

$$f_a = 18{\cdot}93\,\mathrm{Re}^{-0{\cdot}316}\left(\frac{t}{D}\right)^{-0{\cdot}927} \qquad \text{(friction factor)}$$

where,

G_m = Air mass flow rate per unit minimum flow area.
N_r = Number of tube rows.
ρ_a = Density of air.
Re = Reynolds number with fin root diameter as characteristic dimension.
t = Fin thickness.
D = Fin root diameter.

As can be seen from the above equation the fin thickness is an important parameter in determining the performance. Most fins are not parallel sided, but are tapered; thus mean fin thickness not 'strip thickness' must be used in the above equation.

4.2. System Pressure Drop

The total system pressure drop consists of a series of factors which must be added to the bundle pressure drop based on minimum free area.

$$DP_S = DP_B + DP_L + DP_P + DP_C + DP_D + \text{Margin}$$

where,

DP_B = Bundle pressure drop from wind tunnel data or correlation.
DP_L = Louvre pressure drop where appropriate.
DP_P = Additional allowance for plenum expansion and contraction losses.
DP_C = Allowance for fan configuration, inlet obstruction and type of ring also includes plenum maldistribution and internal recirculation.
DP_D = Dirt film allowance. This is an optional addition for plants in fouling areas and is usually quoted as a dirt film thickness which must be added to both sides of the mean fin thickness when calculating pressure drops.

Margin
This is often requested in conjunction with autocontrol so that the fans can operate above as well as at the design point, typically 120%. Thus the total system resistance is likely to be considerably above the bundle pressure drop calculated by correlation. At present, research is being directed on to the losses associated with plenum shape, height and aspect ratio for both forced and induced draft units. Typical plenum heights offered by most air cooler manufacturers can be quantified against tube length.

TABLE 6.1
RELATIONSHIP OF PLENUM HEIGHT TO TUBE LENGTH

Tube length (m)	*Plenum height (m)*
3·0	0·5
3·0–6·00	0·8
6·0–9·00	1·0
9·0–12·00	1·2
12·00–15·00	1·5

The justification for the figures in Table 6.1 is basically economic, however some preliminary research suggests deeper plenums may be the optimum. Current research will hopefully quantify the losses associated with the depths actually used. Pending this information the following allowances are suggested:

(a) Plenum losses 0·5–1·2 fan velocity heads.
(b) Configuration losses 10% of bundle static.

4.3. Fan Drives and Transmissions

The airflow for air-cooled heat exchangers is usually provided by axial flow fans. For units longer than 4 m it is usual to provide two or more fans along the tube length. It is quite usual for several tube bundles on the same service, or even independent units, to share the same fan bay. However, it is strongly recommended that partition walls between fan bays should not occur in the middle of bundles. Shutdown of a fan in this situation could lead to severe temperature gradients across a header box, with the risk that the resulting 'warping' could damage either the header or the bundle.

There are four major types of drive associated with air cooler fans. The prime mover is usually an electric motor but is sometimes a steam turbine, air hydraulic motor.

Vee-belt Transmission

These are either of the multi-stranded type or, more recently, of the banded type. Some specifications such as API 661[3] limit their use to a maximum of 22 kW drives. However, with the speeds associated with 50 Hz four-pole motors, most European companies are happy to utilise belts with suitable profiles up to 37 kW drives.

Toothed-belt Transmission

These are of the 'timing band' type and are more expensive than vee-belts because of the need to cut teeth on the pulleys. Additionally they can cause a noise problem in the region 1000–2000 Hz unless close-cowled with an insulated guard. However toothed-belts are often preferred since they do not need such regular re-tensioning as vee-belts.

Direct Drive

This is normally used on fans of up to 2 m diameter with electric motors having up to eight poles. Motors with more poles are available and have been used on a few installations for larger diameter fans, but usually economics rule them out.

Gearbox Transmission

Gearboxes are used in conjunction with electric or steam turbine drives where the power transmitted is too high for the other types of transmission discussed above. Many types of gearboxes have been used although bevel and parallel gears have proved most satisfactory.

API 661[3] recommends a service factor of 2·0 for air cooler drives. Some gearboxes can be damaged by backwards rotation which can occur when the fan windmills under natural draft. In this situation mechanical or electromagnetic anti-rotation devices must be considered.

4.4. Effect of Noise on Fan Selection

Probably the most important parameter affecting the selection of air cooler fans is noise. Air cooler fans can typically contribute 50% to the noise emission of a large refinery or chemical plant.

The most obtrusive noise level is in the region 250–1000 Hz. However, at plant boundaries, these frequencies are dominated by lower frequencies which are not absorbed as fast with increasing distance. The noise of air coolers at a plant boundary can be described as a 'low rumble'. Thus two types of noise limit emerge:

Plant Noise

This is usually set by national legislation. Most countries limit noise at the nearest walkway or access point to 85 dB(A), and this is usually interpreted as at 1 m from the fan.

Community Noise

As discussed above, in order to limit the noise nuisance at the plant boundary, limits are increasingly being set on plant noise. As it is difficult to measure noise on a crowded site at a distant point due to a particular piece of equipment, guarantees are normally set on sound power level (PWL). Sound power level can be measured by reading close to the equipment[5] and standard tables used to extrapolate to sound pressure levels at a particular distance. Noise varies with:

(a) Make of fan.
(b) Inlet configuration.
(c) Plenum configuration.
(d) Type of drive.
(e) Fan tip speed.

Fan tip speed is the most important parameter affecting overall noise, although inlet configuration has a considerable effect on low frequency noise. Typically, for fans in the 3–4·5 m range with 22 kW drives, the noise levels indicated in Table 6.2 can be achieved.

Lower noise levels can only be achieved by increasing the surface, reducing fan power and allowing further reductions in tip speed. At low tip speeds air cooler designers should ensure that the performance of the fan has been adequately tested. Blade theory can only be applied for fans of low

TABLE 6.2
TYPICAL FAN NOISE LEVELS

Type	*Tip speed (m/s)*	*SPL at 1 m (dB(A))*	*Sound power level/fan (dB(A))*
4 blades	60	88	100
4 blades	55	85	98
6 blades	45	80	93
6–12 blades	30	73	86
8–24 blades	25	70	83

(The number of blades depends on the blade solidity.)

solidity with up to six blades. Above this figure interference begins to occur between adjacent blades.

Good low-noise fans tend to have large hubs occupying up to 50 % of the diameter with high solidity blades set in a smooth bell-shaped inlet in spun steel or plastic. The blades—because of the high solidity—are either plastic or extruded/fabricated aluminium.

4.5. Fan Aspect Ratio

This is defined as the ratio of bundle face area to fan area. The bundle face area is calculated excluding bundle side frames but the fan area is usually calculated including the hub. The normal limit is set at a maximum aspect ratio of 2·5 although a higher limit could be argued for induced draft units. For low-noise fans with large hub areas a lower limit is preferable. Most air cooler fan bays are rectangular rather than square. Thus a subsidiary limit is usually set, in that the fans should cover at least 55 % of the tube length.

5. DESIGN AND CONSTRUCTION

5.1. Thermal Design

Air coolers are designed to the classic equation of heat transfer,

$$Q = U_f A_f \theta_m \tag{1}$$

where,

Q = Total duty (W)
U_f = Overall finned surface heat transfer (W/m^2 h °C)
A_f = Extended fin surface (m^2)
θ_m = Effective mean temperature difference (°C)

The overall heat transfer coefficient U_f (eqn. (1)) can be further defined in terms of the outside fin coefficient, inside coefficient, fouling factor and tube wall resistance.

$$\frac{1}{U_f} = \frac{1}{h_e} = R_c + \frac{A_f}{A_I}\left[\frac{1}{h_i} + R_f + \frac{t_w}{k_t}\right] \tag{2}$$

where,

h_e = effective extended surface heat transfer coefficient (W/m^2 h °C).
A_f, A_I = ratio extended to inside tube surface.
h_i = inside heat transfer coefficient.
R_f = in-tube fouling resistance.
R_c = fin contact resistance.
k_t = tube wall conductivity.
t_w = tube wall thickness.

The extended surface heat transfer coefficient can be calculated from the Briggs and Young correlation.[6]

$$h_a = 0{\cdot}134\,\mathrm{Re}^{0\cdot681}\,\mathrm{Pr}^{0\cdot33}\left(\frac{P}{l}\right)^{0\cdot2}\left(\frac{P}{t_f}\right)^{0\cdot1134}\frac{k_a}{D_o}$$

where,

Pr = Prandtl number air.
t_f = fin thickness (m).
P = fin pitch (m).
l = fin height (m).
D_o = core tube outside diameter.
k_a = thermal conductivity of air.
h_a = extended surface heat transfer coefficient.

This coefficient has to be modified to arrive at the effective outside heat transfer coefficient

$$h_e = h_a\left[1 - (1 - \eta_f)\frac{A_f}{A_o}\right]$$

where,

η_f = fin efficiency,
A_o = outside bare surface (m^2).

The fin efficiency is often calculated by a method proposed by Kraus for parallel plates.

Contract resistance R_c is normally assumed as zero for mechanically embedded fins (G fin) and for tubes where the bond is provided by galvanising. For other types of fin, such as L fins and extruded fin tubes where the bond is provided by tension, the gap resistance increases with temperature when dissimilar tube and fin materials are used. In the case of a tension-wound aluminium L fin on a carbon-steel core tube, the effective heat transfer coefficient can be reduced by more than 10% for a temperature rise of 100 °C above the finning temperature.

Initially the designer knows only the chosen airflow and the inlet air temperature, the design method is iterative with a backwards integration technique used on a series of steps from the outlet to the inlet, the outlet air temperature from a particular row of tubes becoming the inlet air temperature for the next. Together with the increasingly complicated methods used to calculate in-tube coefficients to produce an accurate solution, the designer must by necessity turn to computer techniques.

5.2. Tube Bundle Construction

An air cooler tube bundle consists of a series of finned tubes set between side frames, supported at a maximum pitch of 1·8 m, passing between header boxes at either end. The layers of tubes are organised into a series of process side passes by internal baffling of the header boxes. For horizontal bundles, vertical baffles result in a crossflow and horizontal baffles a counterflow arrangement relative to airflow. Nozzles are attached to the inlet and outlet which may or may not be on the same header box. The header boxes at each end may need to be 'split' into double boxes if the potential temperature difference on the process side exceeds 150 °C, this is to prevent twisting of the headers with the possibility of damaging the tube attachment. Two types of tube attachment are generally available:

(a) Rolling into grooved tubesheets.
(b) Welding into tubesheets (with or without filler wire).

Many types of header boxes are offered by industry, however, they can be grouped into three general classifications:

(1) Plug type boxes.
(2) Removable cover plate boxes.
(3) All-welded construction.

5.3. Plug Headers

Here access plugs are fitted opposite the tube ends to enable the tubes to be

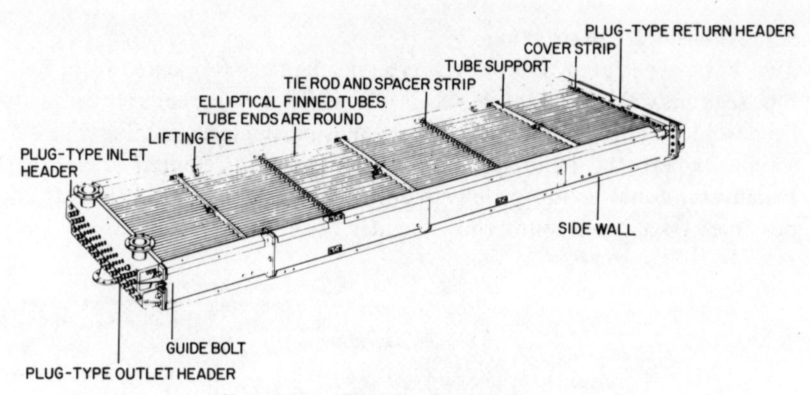

FIG. 6.10. Plug header bundle.

fixed to the tubesheet. Limited access is available for cleaning because the method of construction allows the plates to be fitted between the tube and plugsheet and the spans are reduced. The plug header gives an economic solution up to pressures of 200 bar g (Fig. 6.10).

5.4. Removable Cover Plates

With cover plates, internal bridging between the tubesheet and removable cover is not possible. Thus mechanically, a cover plate header must be calculated with the full height of the box as the characteristic dimension. This makes cover plates more expensive, and pressure limits them to below 30 bar g. Access is easily available for cleaning and cover plates should be used in high fouling service where possible. Various forms of cover plate header are offered with studded or flanged and through-bolted attachments (Fig. 6.11).

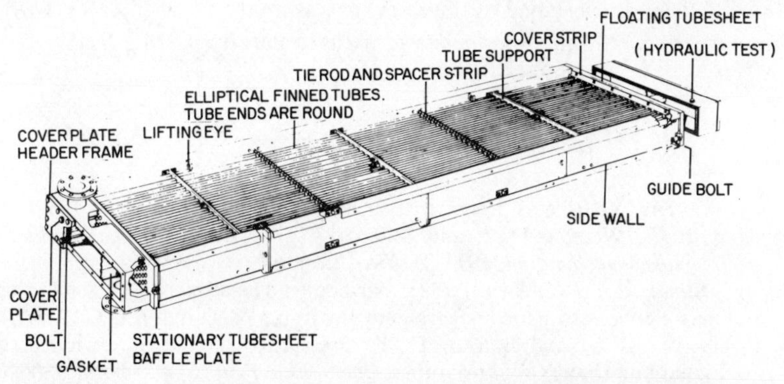

FIG. 6.11. Removable cover-plate bundle.

5.5. All-welded Construction

Two basic types are available. 'D' type headers are fabricated from half-pipe sections welded to flat tubesheets into which the tube ends have already been welded. This method of construction is particularly suitable for clean services where the minimum leakage potential is required (Fig. 6.12). Because of construction problems only crossflow pass arrangements are possible. To achieve counterflow, header boxes must be split.

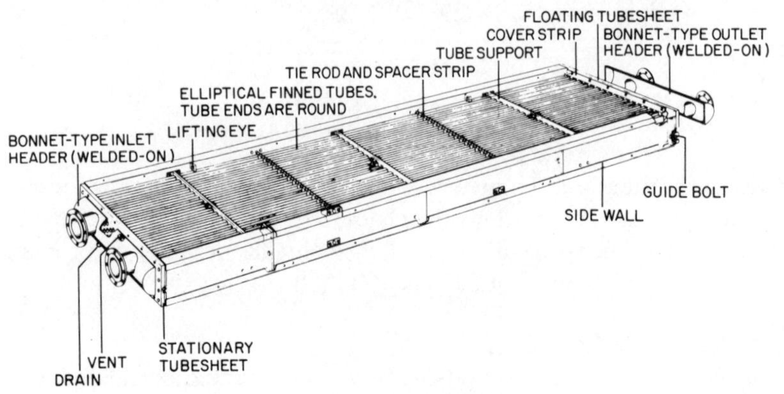

FIG. 6.12. All-welded bundle.

For higher pressures, welded manifold construction is required, individual tubes are welded into serpentines with 'U' return bends which in turn are welded into pipe inlet and outlet manifolds. This construction is very expensive but is suitable for use up to 600 bar g. Many variations exist on header box types, but in general they fall into the three groups discussed above. All can be designed to the relevant national codes; however, within the petrochemical industry the standard is generally ASME VIII.

REFERENCES

1. GANAPATHY, V. Process design criteria, *Chem. Eng.*, March 1978, 108–11.
2. ROSE, J. C., WILSON, D. J. and COWAN, G. A. *Air-cooled heat exchanger performance specification*, 1973, HMSO, London.
3. *API Standard 661* (2nd ed.), 1973, Air-cooled Heat Exchanger for General Refinery Services, American Petroleum Institute, Washington D.C., USA.
4. ROBINSON, K. K. and BRIGGS, D. E. Pressure drop of air flowing across triangular pitch banks of finned tubes, *Chem. Eng. Prog. Sym. Ser. 62*, 1966, **64,** 177–82.

5. Concawe Noise Advisory Group—Task Force No. 2. *Test Method for Determining the Sound Power Levels of Air-cooled (Air-fin) Heat Exchangers*, December 1978, Von Hoghenhouckloon, 60 2596 TE, Den Haag, Netherlands.
6. Briggs, D. E. and Young, E. H. Convection heat transfer and pressure drop of air flowing across triangular pitch banks of finned tubes, *Chem. Eng. Prog. Sym. Ser. 59*, 1963, **41,** 1–10.

Chapter 7

SPECIAL SURFACE GEOMETRIES FOR HEAT TRANSFER AUGMENTATION

RALPH L. WEBB

Department of Mechanical Engineering,
The Pennsylvania State University, University Park, Pennsylvania 16802

SUMMARY

This chapter discusses special surface geometries, which may be used to enhance heat transfer in the various types of heat exchangers used in practice. The applicable modes of heat transfer include single-phase convection of liquids and gases and boiling and condensing fluids. Although the general technology of enhanced heat transfer includes 'active' techniques such as electric and acoustic fields, we have excluded those in favor of special surface geometries, which promise greater commercial interest. The survey includes information from the US Patent literature, as well as journal literature. This chapter discusses 93 references, selected from more than 2300 available. We attempt to discuss the merits of the several competing techniques in an attempt to identify those of greatest potential. The practical selection of enhanced surfaces requires more than knowledge of their basic heat transfer and friction characteristics. We discuss guidelines to define when augmented surfaces may be used to advantage. Having established that potential interest exists, we address Performance Evaluation Criteria (PEC) which establish the quantitative advantages for specified design constraints. These PEC are discussed for four basic design objectives. Further, different PEC may be required for single- and two-phase exchangers. Finally, we speculate on the future of augmentation technology, considering expected and needed developments.

NOMENCLATURE

A	heat transfer surface area
D	inside diameter of tube
e	height of fin or roughness element
f	friction factor
G	mass velocity in tube (kg/s)
n	number of internal fins in tube of diameter D
L	height of vertical condensing surface
LMTD	log mean temperature difference
p	axial spacing
P	pumping power
ΔP	pressure drop
Pr	Prandtl number
Q	heat transfer rate (watts)
Re	Reynolds number
r	thermal resistance ratio (outside-to-inside) of reference smooth-tube heat exchanger design
U	overall heat transfer coefficient
V	volume of heat exchanger tubing material

Subscripts

s	refers to reference smooth tube exchanger design

Greek symbols

α	heat transfer coefficient
α_{Nu}	condensation coefficient on vertical surface predicted by Nusselt theory
β	helix angle of internal fins or internal roughness, measured from tube centerline
η	efficiency index $(\alpha/\alpha_s)/(f/f_s)$
σ	surface tension

1. INTRODUCTION

This chapter is concerned with the application of augmentation techniques to heat exchangers. Bergles and Webb[1] have defined 13 techniques to provide heat transfer augmentation. These techniques are categorized into two groupings: passive and active techniques. Passive techniques employ special surface geometries or fluid additives for enhancement. The active techniques require external power, such as electric or acoustic fields, or surface vibration. Potentially, the 13 techniques may be considered for application to several heat transfer modes, which include single-phase free and forced convection, boiling and condensation. Reference 2 presents 1967 literature citations classified by heat transfer mode and augmentation technique. The citation distribution is given in Table 7.1.

In addition to the journal literature, the patent literature presents several hundred additional augmentation references. This patent literature may be particularly interesting to designers, since it contains ideas conceived by industry and of commercial interest.

The majority of commercially interesting augmentation techniques are limited to the passive techniques, which employ special surface geometries. The special surface geometries provide augmentation by establishing a higher αA per unit of projected surface area. Three basic methods are employed to obtain increased αA: (1) increased heat transfer coefficient (α) without increased surface area, (2) increased surface area (A) without increased heat transfer coefficient, and (3) extended surface areas which provide both increased surface area and higher heat transfer coefficients on the extended surface. Concepts which yield an increased heat transfer coefficient without area increase include surface roughness and turbulence promoters. Surface area increase is provided by extended or finned surfaces. Finned or specially formed surfaces may also provide higher heat transfer coefficients due to reduced hydraulic diameter or the establishment of reduced boundary layer thickness. Such examples include interrupted fin geometries for single-phase forced convection, short fins for condensation on horizontal tubes, and the use of surface tension forces to thin the film thickness in condensation.

The majority of two-fluid heat exchanger applications are served by four basic exchanger types: (1) shell-and-tube or tube banks, (2) fin-and-tube, (3) brazed-plate fin, and (4) plate-type exchangers. These exchanger types are commonly employed for single-phase forced convection to liquids and gases, boiling and condensation. If enhancement is employed, it must be applied to three basic flow geometries:

1. Internal flow in circular tubes.
2. External flow across and along circular tubes.
3. Channel flow in closely spaced parallel plate channels typical of plate fin and plate-type exchangers.

A variety of materials may be employed for fabrication of the heat

TABLE 7.1

CLASSIFICATION AND DISTRIBUTION OF AUGMENTATION TECHNIQUES

	Single-phase natural convection	*Single-phase forced convection*	*Pool boiling*	*Forced-convection boiling*	*Condensation*	*Mass transfer*
Passive techniques (no external power required)						
Treated surfaces	—	—	65	14	24	—
Rough surfaces	2	200	15	32	15	14
Extended surfaces	*	210	28	31	75	10
Displaced enhancement devices	—	28	—	7	1	3
Swirl flow devices	—	88	—	72	10	4
Surface tension devices	—	—	6	—	—	—
Additives for liquids	3	10	41	16	—	1
Additives for gases	—	132	—	—	5	2
Active techniques (external power required)						
Mechanical aids	4	13	14	+	13	8
Surface vibration	44	22	9	2	5	5
Fluid vibration	43	80	16	4	1	19
Electric or magnetic fields	37	30	19	8	11	6
Injection or suction	2	10	6	+	6	+
Compound techniques	+	10	+	+	1	+

— Not applicable.
* Not considered in this survey.
+ No citations located.

transfer surface. Typical of these are copper, aluminum, steel and titanium. Corrosion risk or high fluid temperatures and pressures may require the use of alloys of brass, bronze or alloy steels. Although an augmentation technique may be available in one material, it will not generally be available in all materials. This is because different manufacturing methods may be required for the different materials.

For a given heat transfer mode, all of the augmentation techniques may be considered to be in competition with one another. Therefore, we would not expect to find uniform acceptance of all techniques. The favored method will provide the highest performance and reliability at minimum cost.

Table 7.2 summarizes the present and potential use of augmented surface geometries for the applications of interest. The first half of the table considers special surface geometries on the inner and outer surface of circular tubes. The latter half of the table considers plate-type and plate fin exchangers. A coding system defines the degree to which an augmentation technique has found application. A dash entry indicates 'not applicable'. The third column lists materials which have been used for commercial manufacture. The last column suggests the performance potential, independent of whether a commercial manufacturing technique exists.

A limited number of the enhancement techniques are commercially used. In only a few cases are techniques considered to be in common use. The next section presents a detailed assessment of the various special geometries.

2. AUGMENTATION TECHNIQUES

This section surveys the technology of heat transfer augmentation by heat transfer mode and flow geometry.

2.1. Forced Convection in Tubes

Wire coil and twisted tape inserts have been commercially employed for some years. However, their use has been limited to special applications. In turbulent flow, they normally require a relatively high friction expenditure compared to the enhancement obtained. Typical heat transfer coefficient increases are 30–100% for wire coil inserts[3] and up to 90% for twisted tapes.[4] The flow obstructions imposed by wire coil inserts offer potential fouling and tube wall erosion problems. Twisted tapes offer higher heat transfer performance per unit of friction in laminar flow.[5]

In turbulent flow, internal roughness and internal fins can provide a given

TABLE 7.2

PRESENT AND POTENTIAL APPLICATION OF AUGMENTED SURFACE GEOMETRIES

	Commercial availability	*Forced convection*	*Mode*			*Performance potential*
			Boiling	*Condensation*	*Typical material*	
Inside tubes						
Metal coatings	Yes	—	2	—	Al, Cu, St	High
Integral fins	Yes	2	3	4	Al, Cu	High
Flutes	Yes	4	4	4	Al, Cu	Moderate
Integral roughness	Yes	2	3	4	Cu, St	High
Wire coil inserts	Yes	3	4	4	Any	Moderate
Displaced promoters	Yes	2	4	4	Any	Moderate (Laminar)
Twisted tape inserts	Yes	2	3	4	Any	Moderate
Outside circular tubes						
Coatings						
metal	Yes	—	2	4	Al, Cu, St	High (Boil)
non-metal	No	—	4	4	'Teflon'	Moderate
Roughness (integral)	Yes	3	2	4	Al, Cu	High (Boil)
Roughness (attached)	Yes	3	4	—	Any	Moderate (For Convection)
Axial fins	Yes	1	4	4	Al, St	High (For Convection)

Transverse fins						
gases	Yes	1	—	—	Al, Cu, St	High
liquids/two-phase	Yes	1	1	1	Any	High
Flutes						
integral	Yes	—	—	2	Al, Cu	High
non-integral	Yes	—	—	4	Any	High
Plate-fin heat exchanger						
Metal coatings	Yes	—	3	—	Al	High
Surface roughness	Yes	4	3	4	Al	High (Boil)
Configured or interrupted fins	Yes	1	2	2	Al, St	High
Flutes	No	—	—	4	Al	Moderate
Plate type heat exchanger						
Metal coatings	No	—	4	—	St	Low
Surface roughness	No	4	4	4	St	Low
Configured channel	Yes	1	3	3	St	High (For Convection)

Use Code:
1 Common use.
2 Limited use.
3 Some special cases.
4 Essentially no use.

TABLE 7.3

PERFORMANCE COMPARISON OF INTERNAL FINS AND TWO-DIMENSIONAL RIB ROUGHNESS

Augmentation geometry	*Material reduction*				*Increased heat transfer*				*Reduced pumping power*			
	V/V_s	G/G_s	e (*mm*)	n	UA/U_sA_s	G/G_s	e (*mm*)	n	P/P_s	G/G_s	e (*mm*)	n
Two-dimensional rib roughness ($p/e = 10$)	0·54*	1·30	0·19	—	1·28**	1·40	0·12	—	0·64	1·40	0·06	—
Internal fins $\beta = 0°$	0·88*	1·22	1·5	25	1·22**	1·40	2	16	0·58	1·40	2·0	5
Internal fins $\beta = 20°$	0·69*	1·15	1·5	25	1·36**	1·41	2	16	0·31	1·77	2·0	5

$Re_s = 25\,000$; $Pr = 3$; tube inside diameter 17·78 mm with 0·71 mm wall thickness (* $r = 0$, ** $r = 0·5$).

enhancement level with smaller pressure loss than given by wire coil and twisted tape inserts. Webb[6] presents data and design recommendations for two important internal roughness geometries. Internally finned tubes give performances comparable to that of internal roughness for moderate Prandtl numbers in turbulent flow.[7] The internal fin performance is dramatically improved if the fins are provided at a helical angle.[7,10] A recent study[11] has also shown that two-dimensional rib roughness applied at 45–50° helical angle is preferable to transverse rib roughness. Because many geometrical variations of an internal roughness type or internal fin are possible, it is important to select the specific geometry which yields the desired heat transfer performance at minimum friction expenditure. Webb *et al.*[8] provide specific recommendations for the selection of internal roughness and internal fin geometries.[7] The selection of such optimum geometric variables requires the use of quantitative performance evaluation criteria, which are discussed in Section 3.

Table 7.3 compares the performance of internally roughened and internally finned tubes for three cases of interest. The comparisons are for Re = 25 000 and Pr = 3 using the preferred surface geometries established in references 7 and 8. The performance ratios (V/V_s, UA/U_sA_s, and P/P_s) are stated relative to a smooth tube exchanger (subscript 's') designed for the same operating constraints and flow rate. The operating constraints are defined in Table 7.4. The G/G_s columns of Table 7.3 show the velocity ratio required to meet the constraints. Since $G/G_s < 1$, the augmented surface geometries require larger flow frontal area than the reference smooth tube design. Internal fins may offer an advantage over roughness from a fouling

TABLE 7.4
POSSIBLE PERFORMANCE EVALUATION CRITERIA

Case	*Basic geometry*	*Flow rate*	*Constraints: fixed*				*Objective*		
			ΔP	P	Q	A	*Increased heat duty*	*Reduced pumping power*	*Material reduction*
FG-1	X	X				X	X		
FG-2	X		X			X	X		
FG-3	X			X		X	X		
FG-4	X				X	X		X	
VG-1		X	X	X		X	X		
VG-2		X			X	X		X	
VG-3		X	X	X	X				X

point of view, although no data exist to support this supposition. This is because scale deposits will be spread over a larger surface area than for a rough surface. Due to the area increase, the net penalty for fouling will be smaller.

Surface roughness offers very attractive turbulent flow performance with high Prandtl number fluids for which the heat transfer coefficient increase is greater than the friction factor increase.[6] Small scale roughness is not effective in laminar flow. In laminar flow, twisted tape inserts and internally finned tubes appear to be the favored augmentation techniques. Marner and Bergles[12] showed that internal fins yield higher performance than twisted tapes.

Of the various augmentation techniques, roughness and internal fins offer the greatest performance potential per unit material, and per unit pumping power. However, high manufacturing cost and low materials availability have limited their use. The internal fin geometries described by Carnavos[10] are commercially available; these tubes are made of copper. Internal fins are less expensive to make in aluminum using a hot extrusion process. However, there are limited applications for aluminum tubes. A boiler manufacturer[14] has developed a spiralled internally rifled tube in steel for boiler super-heater applications. Nishikawa *et al.*[88] provide test data on steel ('single rifled' and 'cross rifled') tubing for boiler applications. The 'cross rifled' tubing significantly delays the critical heat flux (CHF) up to high steam qualities and provides increased post CHF heat transfer coefficients.

A US Patent[13] describes a recent commercial entry having internal two-dimensional ribs applied at a helical angle. This copper tube also has low integral fins on its external surface.

Within the category of surface roughness, a three-dimensional roughness, e.g. 'sand-grain' type, appears to offer the highest performance.[6] Two US Patents[9,15] describe possible methods for commercial manufacture of sand-grain type roughness inside tubes. The favored roughness of Fenner and Ragi[9] consists of a single layer of metal particles covering approximately 50% of the projected surface. Their tests with water (Re = 35 000, Pr = 10) with $e/D = 0{\cdot}012$ provided 110% enhancement with a 78% efficiency index (η). This efficiency index is lower than that reported by Dipprey and Sabersky in reference 6.

Recent data,[89] with water (Pr = 5) in spirally corrugated tubes (see Fig. 7.7(a)) show $1{\cdot}1 < \eta < 1{\cdot}2$ for $4000 < \mathrm{Re} < 16\,000$ with a 'severity factor' $e^2/pD \simeq 0{\cdot}002$. At this severity factor and $p/e = 10$, $e/D = 0{\cdot}02$, Gupta and Rao,[89] obtained $St/St_s = 1{\cdot}75$.

2.2. Forced Convection in Tube Banks

Extended surfaces have become standard for single phase forced convection to gases, and are commonly employed with liquids. The main flow geometry of interest is cross flow, although there are applications for axial flow.

2.2.1. Crossflow

Finned-tube banks for gases use either the plate fin or circular fin geometry. A wavy or interrupted slit fin geometry is employed to obtain higher heat transfer coefficients with plate fins. The circular fin may be slit along radial lines to provide a high performance interrupted fin geometry. Air conditioning applications use 5–8 fins/cm, while process applications are usually limited to 4 fins/cm. Soot-laden gases may limit the fin density to 2–4 fins/cm. Staggered tube banks yield higher performance than inline tube banks.[16] The strong by-pass stream in inline tube banks is responsible for their lower performance. The performance of a finned-tube bank may be improved using flattened or oval tubes. This reduces the large parasitic drag loss due to the tubes. The performance of several varieties of oval and flattened tubes is given in references 17 and 18.

Webb[19] presents a detailed survey of air-side extended surface geometries and the performance characteristics of finned-tube banks. This survey contains 88 references to further information, most of which deal with enhanced fin geometries and high performance tube-bank designs.

Extended surfaces are also used for liquids in crossflow, where shell side fouling is not severe. The fin height must be kept small, e.g. 1·5–2 mm, due to the high heat transfer coefficients of most liquids. Popular commercial geometries have 4–10 fins/cm which are formed from a thick walled tube using a thread rolling process. These tubes are available in virtually any material.

Roughness on the tube outer surface has not been used for single-phase heat transfer in shell-and-tube exchangers. Groehn and Scholz[20] tested knurl-roughened tubes ($e/D = 0{\cdot}03$ and $0{\cdot}017$) with an inline tube bank in crossflow. The heat transfer coefficient was improved 50%, while the pressure drop was slightly reduced. Because roughness causes earlier transition to turbulence, the boundary layer separation is delayed to a greater circumferential position, which acts to reduce the drag coefficient.

2.2.2. Axial Flow

Axial flow in rod bundles is of interest for gas-cooled nuclear reactors. Heat transfer is augmented using small transverse ribs spaced 10–20 times their height. The ribs may be integral with the tube wall or in the form of a helical wire wrap.[21,22] White and Wilkie[21] have shown that a helical angle of approximately 45° is preferred. Using the hydraulic diameter concept, data on roughness inside circular tubes should agree with that for axial flow in rod bundles.[23]

Axial fins may be applied to the outer surface of a tube. This geometry is of interest for double-pipe heat exchangers and tank heaters, where the heat transfer fluid is gas or an oil. Axial fins are commercially available as aluminum extrusions or welded steel fins.

2.3. Boiling

Enhancement is of interest for nucleate boiling and thin film evaporation on the outer surface of a tube bundle, and for flow boiling inside tubes.

2.3.1. Nucleate Boiling Outside Tubes

Low, integral-fin tubing has been used for many years. However, rapid advances have been made in the development and commercial application of special nucleate boiling geometries. These geometries establish nucleate boiling at small surface–liquid temperature differences, and yield considerably higher performance than those obtainable with integral-fin tubing. The first of these special geometries was commercially available in 1969.[24,25] Today, six nucleate boiling geometries are commercially available, or reported in the US Patent literature. Four of these have been reported after 1973. Two basic approaches are employed to form a high area density of nucleation sites:

1. *Porous boiling surfaces:* The Linde Division of Union Carbide Company pioneered the porous boiling surface.[25] Figure 7.1(a) illustrates the cross-section of such a surface and 7.1(b) shows the performance improvement provided for several fluids. It consists of a sintered, porous metallic matrix, which is bonded to the base surface. The porous coating is approximately 0·25 mm thick, has a 50–65% void fraction, and may be made in several materials. The most important dimension is the average pore radius, which ranges from 0·01–0·1 mm, measured by a capillary rise test. Smaller pore sizes are preferred for low surface-tension fluids. Gottzman *et al.*, and Czikk and O'Neill[26,90] have developed a theory to predict the performance of the porous surface, and to establish the preferred pore size. An alternative method for making a porous boiling surface was proposed by Dahl and Erb[28] in a US Patent. The porous coating is formed by flame-spraying aluminum particles in an oxygen-rich

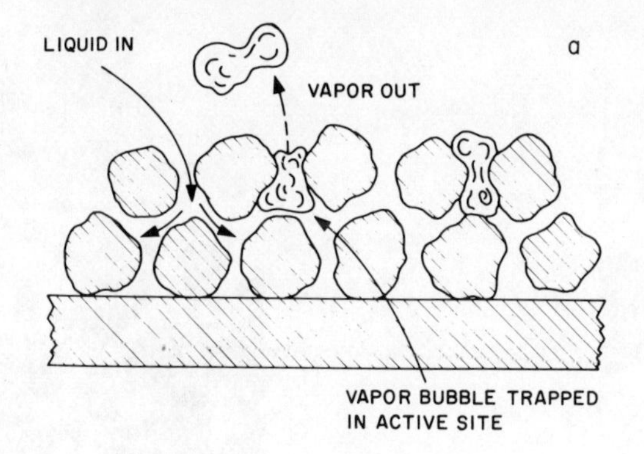

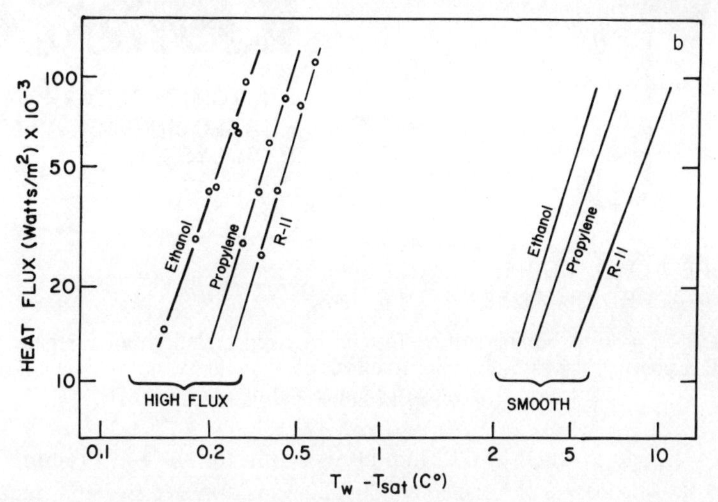

FIG. 7.1. Linde porous boiling surface.[26] (a) Illustration of section through porous coating, (b) boiling performance data.

atmosphere. The oxygen-rich flame produces an oxide film on the particle and prevents the particle from flattening on impact with the surface.

Janowski *et al.*[29] have developed an innovative process for making a porous boiling surface shown in Fig. 7.2. A tube is wrapped with an open-cell polyurethane foam, which is then copper plated. The copper plating provides structural integrity and good thermal conduction. The polyurethane is removed by pyrolysis, which forms additional very small pores within the skeletal structure. The structure described in a US Patent has 1·5-mm coating thickness, 97% void volume, 4 pores/mm and 0·12 mm

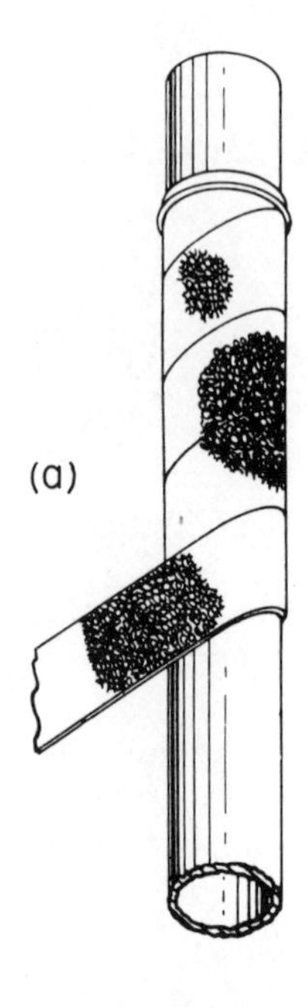

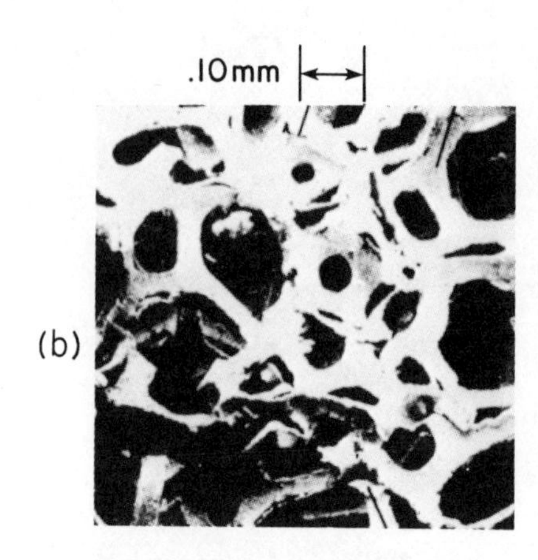

FIG. 7.2. Porous boiling surface formed by copper plating polyurethane foam surface coating.[13] (a) 1·5-mm thick foam wrapped on tube; (b) 100 × magnification of foam structure after upper slating and pyrolysis.

pore size, augmented by 0·02 mm pores within the skeletal structure. The pyrolysis process is performed at 300–520 °C as opposed to 960 °C required by the Linde sintering process. Janowski claims the lower processing temperature is advantageous because the mechanical properties of the base tube are not affected. The flame spray process does not affect the base tube material properties.

2. *Integral 'roughness':* The metal is cold-worked to form a high area density of re-entrant nucleation sites which are interconnected below the surface. Figure 7.3 shows the cross-section of three commercially used boiling surfaces of this type.[30,31,32] Figure 7.3(a) from Webb's US Patent[30] consists of integral fin tubing (13 fins/cm, 0·8 mm fin height) in which the fins are bent to form re-entrant cavities. The groove opening at the surface is a critical dimension, and the recommended gap spacing is 0·0038–0·0089 mm for Refrigerant-11. The performance is sharply reduced

for gap widths outside of this range. The preferred gap spacing is expected to be smaller for low surface-tension fluids.

A variation of the Webb patent[30] developed by Fujii *et al.* of Hitachi Corp.[31] is shown in Fig. 7.3(b). It is formed from a low fin tube, which has small spaced cutouts at the fin tips. These 'saw tooth' fins are bent to a horizontal position to form tunnels having spaced pores at their top. This commercially available tube is called 'THERMOEXCEL-E' and is further

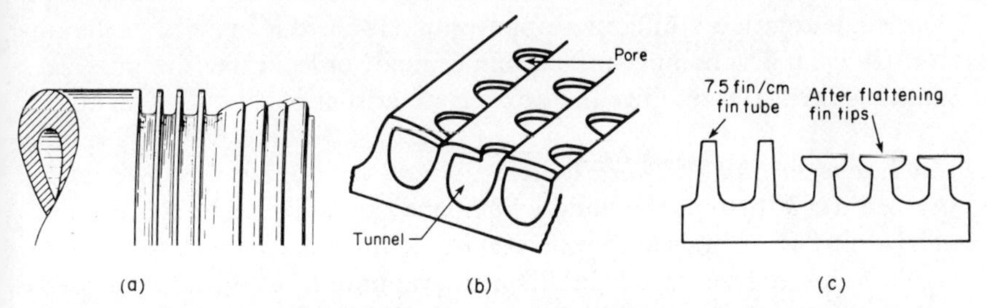

FIG. 7.3. Cross-sectional views of three commercial 'rough' boiling surfaces. (a) Trane bent fins;[30] (b) Hitachi bent 'saw-tooth' fins;[31] (c) Weiland flattened fins.[34]

discussed by Arai *et al.*[32] and Torii *et al.*[33] Nakayama *et al.*[91] propose a model for boiling in the enclosed tunnels, and develop an analytical model for correlation of the boiling performance as a function of the geometric and fluid properties. The Webb and Hitachi surfaces were developed for use in large refrigeration machines and employ copper tubes. In 1979, the Weiland Werke A.G.[34] announced a third nucleate boiling roughness geometry, illustrated by Fig. 7.3(c). It is basically a variation of Fig. 7.3(a) and may be formed from standard 7·5 fin/cm integral-fin tubing. The uniformly smooth outer surface of this tube geometry does not require the formation of smooth tube 'land' sections for mating at tube supports. This provides higher performance over the entire tube length.

Kun and Czikk[35] describe a similar re-entrant grooved surface. A series of fine, closely spaced grooves (11–90 grooves/cm) are cut in the surface, followed by a second set of cross grooves, which are not as deep as the first set. The US Patent provides data on boiling nitrogen and water for a variety of groove densities. Kun also reports that the opening width of the re-entrant cavity is a critical dimension, as reported by Webb. Maximum performance is obtained with a 'restricted opening' dimension 0·0013 mm for nitrogen and 0·006 mm for water.

The performance exhibited by the several types of coated and rough surfaces is nominally comparable, with perhaps the highest performance given by the Linde sintered coating. The key to the high performance of the different structures is attributed to three factors: (1) a pore or re-entrant cavity within a critical size range, (2) interconnected cavities, and (3) nucleation sites of a re-entrant shape. When the cavities are interconnected, one active cavity can activate adjacent cavities. It appears that the dominant fraction of the vaporization occurs at very thin liquid films within the subsurface structure. The re-entrant cavity shape provides a very stable vapor trap, which will remain active at very low liquid super-heat values. The coated surfaces offer the opportunity for a duplex tube material construction. The boiling surface may be made of less expensive material than required to meet the corrosion characteristics of the tube-side fluid.

2.3.2. Evaporation of Thin Films

As an alternative to pool boiling, a liquid may be vaporized by distributing it as a thin film on the heat transfer surface. This method may be preferred if static head effects, e.g. in a flooded evaporator, cause an increased saturation temperature at the heat transfer surface. Because heated films are susceptible to rupture, the liquid rate must be sufficient to assure surface wetting. Heat exchanger geometries for thin film evaporation include sprayed horizontal bundles and vertical tube evaporators (VTE), with vaporization occurring inside the tube. Enhanced surface geometries for VTE include fluted tubes[10,14,36,37,38] and internally finned tubes.[39] Doubly fluted tubes, Figs. 7.4(b) and (c), were developed for seawater distillation, with augmentation of condensation on the outer surface, and brine evaporation on the inner surface. The augmentation mechanisms on fluted surfaces are different for condensation and evaporation. In condensation, surface tension drains the film from the convex surface of the flutes (Fig. 7.4(a)). In evaporation, the feed tends to be channelled in the valleys, and is believed to wet the convex surface by splashing or film instabilities. Nucleation within the film will promote surface wetting. Johnson *et al.*[82] propose possible enhancement mechanisms. Carnavos[72] reports a three-fold increase in evaporation coefficient (based on total surface area), and Alexander and Hoffman[36] report overall coefficient increases of 2–3 for doubly fluted tubes evaporating fresh water. Johnson *et al.*[82] found that higher coefficients are obtained with seawater than with deionized water. The nominal 100% higher values are due to interfacial motion caused by surface tension gradients. The addition of a very small amount of surfactant gives a 100% higher overall coefficient with a doubly fluted tube in an upflow VTE.[38] The surfactant causes very thin foamy liquid layers, which wet the entire fluted surface.

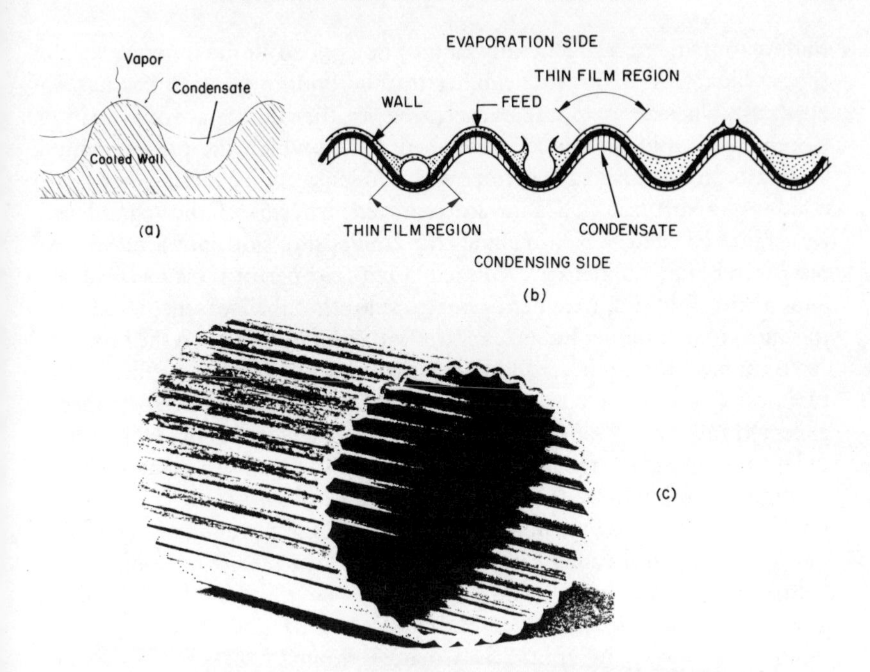

FIG. 7.4. Fluted tubes. (a) Cross-section of vertical fluted condensing surface;[20] (b) cross-section of doubly fluted tube for condensation on the outer surface and thin film evaporation on the inner surface;[72] (c) photo of doubly fluted tube.[72]

Conti[40] tested commercial low fin tubing (7·9 and 13·4 fins/cm) with ammonia in a horizontal sprayed-film evaporator. The best performance was given by the 7·9 fins/cm tubing with 0·50-mm high fins. This tube has an 80% area increase, and provides a 200% evaporating side enhancement, based on plain tube area. Edwards *et al.*[41] and Schultz *et al.*[42] present an analytical model and experimental results for thin film evaporation on triangular threaded horizontal tubes. Their enhancement levels for 3·9 and 6·3 grooves/cm were lower than measured by Conti[40] with 7·9 fins/in integral fins.

Fricke and Czikk[92] report test results on a sprayed horizontal tube evaporator having the Linde porous coating on the outer tube surface. The ammonia heat transfer coefficients were approximately equal to those obtained with the same surface geometry as a flooded evaporator.[68]

2.3.3. Flow Boiling Inside Tubes

The special surface geometries used for enhancement of nucleate pool boiling (outside tubes) generally cannot be applied to the inner surface of tubes. Therefore, their applicability to flow boiling inside tubes has not been established. A notable exception is the Linde porous coating. Commercial applications are described[27,43] in which the porous boiling surface is on the inner tube surface with condensation occurring on a fluted condensing surface. The data are reported in terms of the overall heat transfer coefficient, which includes the condensing side enhancement. An oxygen reboiler–condenser exhibited 'a ten-fold performance advantage', and a 'five-fold improvement over a smooth tube' was achieved in a thermosyphon reboiler for a C_2 splitter with ethane boiling on the tube side. No data have been reported to define the tube-side boiling coefficient as a function of mass velocity and vapor quality. Favorable performance is expected at low vapor qualities and at surface–liquid temperature differences within the nucleate boiling range of the nucleate boiling surface. At high vapor qualities, the liquid film thickness is not sufficient to permit nucleation. Murphy and Bergles[44] found that the porous boiling surface did not yield improvement in high heat-flux saturated flow boiling.

Special internal roughness geometries have been developed for commercial use. Withers and Habdas[45] tested tubes having 'integral internal helical ridging' (Fig. 7.7(a)) using R-12 in a direct expansion water chiller. Specific surface configurations were tested which gave a 200% increased boiling coefficient. Although the pressure drop was substantially increased, which reduced the LMTD, the best geometry provided a 100% increase in evaporator capacity. Ito *et al.*[46] and Lord *et al.*[47] describe other roughened tubes developed for application to direct-expansion refrigerant evaporators. The tube of Ito *et al.* has spiralled triangular thread-type grooves approximately 0·2 mm deep. This tube provides boiling heat transfer coefficients 1·5–2 times higher than a smooth tube, with very small increased pressure drop.

Kubanek and Milletti[48] and Lavin and Young[49] tested several commercially available internally finned tubes using R-22. The tubes were nominally 15-mm inside diameter with 16 fins[49] and 32 fins,[48] and included axial and spiralled fins. Kubanek and Milletti found that fin spirals yielded increased heat transfer coefficients, and the enhancement was equal to or greater than the area increase. The heat transfer coefficients based on total surface area were 80–300% greater than the plain tube values. Lavin and Young measured local heat transfer coefficients and established tube performance as a function of the two-phase flow regime. They found the greatest enhancement (100%) occurred at low vapor qualities, and the axial fin tube yielded heat transfer coefficients 25% smaller than the smooth tube coefficients in the mist flow regime.

Data have also been reported[49,50] for full-height aluminum fins made as an extrusion, which are shrink-fitted in a copper tube. These generally give lower heat transfer coefficients (total area basis) and a higher pressure drop than low internal fins.

Bryan and Seigel[51] tested an R-11 evaporator using wire coil inserts in a smooth tube. Their tests of five different wire coil geometries provided enhancement up to 200 %. Larson *et al.*[52] found similar improvements with R-12.

Heat transfer coefficients in two-phase flow vary with vapor quality, due to the different flow regimes. Therefore, a given augmentation technique may not be equally effective in all flow regimes. One example is the use of twisted tapes or swirl devices to delay the critical heat flux in sub-cooled boiling[54,55] or to delay dryout in the mist flow region.[56] Performance of twisted tape inserts encompassing different flow regions are given by Blatt and Adi[57] for R-11 and by Nooruddin and Murti[58] for air–water mixtures. They report heat transfer coefficient increases in the 80–100 % range.

2.4. Condensation

The heat transfer resistance in condensation of pure components is primarily due to conduction across the condensate film thickness. Any technique which causes a reduced film thickness will provide enhancement. Special surface geometries or surface treatments are effective in attaining this goal. High vapor velocity which promotes interfacial shear or condensate entrainment will also cause higher heat transfer coefficients. Enhancement may be applied to the outer surface of tubes (vapor space condensation) or to the inner tube surface (forced convection condensation). Because the tube orientation will affect its condensate drainage characteristics, one must distinguish between horizontal and vertical tube orientations. Augmentation is of primary interest for organic fluids, whose condensation coefficient is considerably lower than that of steam.

2.4.1. Condensation on Horizontal Tubes

If surface wetting can be prevented, high performance dropwise condensation will occur.[59] Chemical additives or surface coatings are effective in promoting drop-wise condensation of high surface-tension fluids, such as water. However, such methods are ineffective in promoting drop-wise condensation for low surface-energy fluids, e.g. refrigerants.[60]

Low integral-fin tubing has found widespread commercial acceptance for condensation on horizontal tubes. Because thin condensate films are formed on the short fins, the condensation coefficient is higher on the finned surface than on a plain tube of equal diameter. Beatty and Katz present a

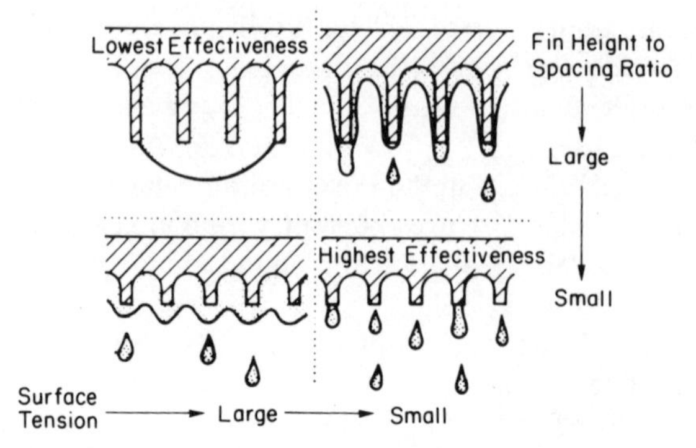

FIG. 7.5. Condensate retention as a function of fin geometry and condensate surface tension.[62]

theoretical model and correlation for this geometry.[61] The permissible fin spacing is limited by the condensate surface tension and vapor shear[62] as shown in Fig. 7.5. If the condensate bridges the spacing between the fins, the extended-surface will not be effective. Fin densities of 7–14 fins/cm (19–35 fins/in) are used for refrigerants or low surface-tension fluids. Externally finned tubes are not normally used for steam condensation due to the probability of condensate bridging and because the limiting thermal resistance is usually on the tube side. However, the use of tube-side augmentation would make steam-side augmentation worthwhile.

Figure 7.6(a) shows a recently developed modification of the integral-fin tube.[32] This commercially available surface, known as 'THERMOEXCEL-C', provides a substantially higher coefficient (based on total area) than the 7·5 fins/cm integral-fin geometry. This tube has 14 fins/cm and 1·2 mm fin height. Data have been reported for Refrigerant-11, -12, -22 and ammonia.[64] Figure 7.6(b) shows a spine-fin array proposed by Webb and Gee.[65] Analytical predictions for condensing R-11 and R-12 show that the spine-fin geometry will provide the same

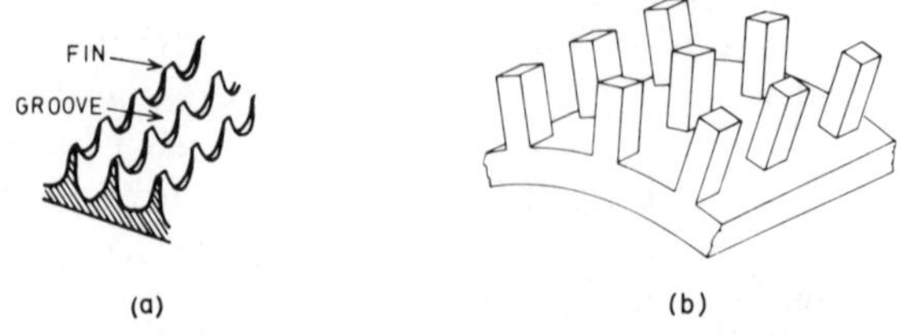

FIG. 7.6. Recent developments in extended surfaces for film condensation on horizontal tubes. (a) Hitachi 'THERMOEXCEL-C';[32] (b) spine-fin surface.[65]

condensing-side performance as integral-fin tubing and permit a 60% reduction of fin material.

Figure 7.7 shows alternate surface geometries for condensation on horizontal tubes. Figure 7.7(a) shows a corrugated tube which has axially spaced helical grooves on the outer wall and inwardly extending helical ridges on the inner wall. Surface tension forces pull the condensate into the helical grooves on the outer surface, and water-side enhancement occurs

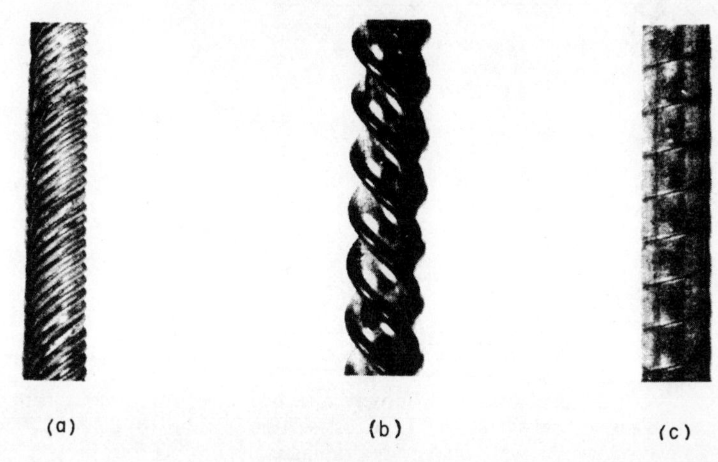

FIG. 7.7. Doubly augmented tube geometries for condensation on horizontal tubes.[67] (a) Helically corrugated tube; (b) helically deformed tube having spiral ridges on outer surface and grooves on inner surface; (c) corrugated tube formed by rolling in helical shape followed by seam welding.

due to the helical ridges on the inner surface. Withers and Young[66] show that this doubly enhanced tube will provide a 30–50% tubing material reduction compared to a smooth-tube steam condenser designed for equal water-side pressure drop. The steam condensing enhancement is of the order of 35–50% relative to a smooth tube. Marto *et al.*[67] performed comparative tests of the geometries in Fig. 7.7 for steam condensation. All geometries provide enhancement on both the water side and the condensing side. Their tests showed that the dominant enhancement occurred on the water side, and that the condensing coefficients on the tubes of Figs. 7.8(a) and (b) are very close to the smooth-tube values. The water-side enhancement was accompanied by a substantially increased pressure drop. These results imply that such tubing would be of value only if the water side is the controlling thermal resistance.

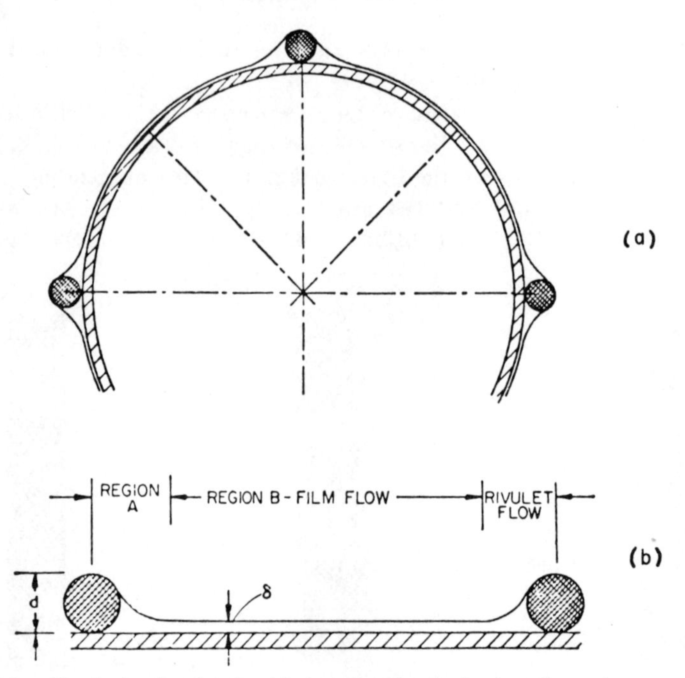

FIG. 7.8. Vertical tube fitted with loosely attached wires for enhancement of condensation on vertical surfaces.[71] (a) Cross-section of tube; (b) detail showing condensate drainage.

Thomas *et al.*[68] have tested a smooth tube helically wrapped with a spaced wire. Surface tension pulls the condensate to the base of the wires, which thus act as condensate run-off channels. Tests involving condensing ammonia on a 38-mm O.D. tube indicated a condensing coefficient about three times the value predicted by the Nusselt equation for condensation on a smooth tube.

2.4.2. Condensation on Vertical Tubes

In 1954, Gregorig[69] proposed a method of using surface-tension forces to enhance laminar film condensation on a vertical surface. Figure 7.4(a) illustrates a horizontal cross-section through the wall of a vertical fluted tube. Due to the surface curvature, the liquid pressure in the convex film is greater than that of the vapor. The combination of convex and concave surfaces establishes a surface-tension force which draws the condensed liquid from the convex surface into the concave region. A high condensation rate occurs on the convex portions of the fluted surface and the concave portions serve as condensate drainage channels. Webb[70] and Panchal and Bell[93] present a general summary of the theory of the fluted condensing surface, define optimum flute geometries and present equations for construction of the surface profile. The resulting heat transfer coefficients averaged over the total surface area are substantially higher than for a uniform film thickness on a smooth tube. The size of the flutes is selected such that the drainage channel will be filled to capacity at the bottom of the vertical surface. Therefore, as the tube length is varied, the flute size should be changed accordingly.

For a given flute geometry, the theory[70] shows that $\alpha_{Nu} \propto L^{-1/4}$. Therefore, $\alpha/\alpha_{Nu} \propto (\sigma/L)^{1/4}$. This shows that the fluted surface is best applied for condensation of higher surface-tension fluids (e.g. water) on long vertical tubes.

Fluted tubes have received considerable attention for vertical tube condensers used in desalination and are commercially available. Thomas[71] and Carnavos[72] give performance data for singly and doubly fluted tubes (see Fig. 7.4(c)). They show augmentation ratios α/α_s (total area basis) in the range of 4–8 for tube lengths between 0·30 and 0·60 m length. Combs and Murphy[73] report similar augmentation ratios for condensing ammonia on 1·2-m high tubes.

A similar effect to that produced by fluted surfaces can be obtained by loosely attached, spaced vertical wires on a vertical surface, as shown in Fig. 7.8. If the wire diameter (d) is appreciably larger than the condensate film thickness (δ) and the wires are wetted by the condensate, surface tension draws the condensate into a rivulet at the wire (region A). This produces film thinning in the space between the wires (region B). The enhancement occurs due to the thinned film in region B, and the condensate drains along the wires. Thomas[71] gives an analytical model and experimental data for this augmentation technique. In a later publication,[74] he shows that square wires are more effective than circular wires of the same diameter. This is because square wires have a greater condensate carrying capacity. Soviet researchers[75,76] have also worked with loosely attached vertical wires. Their work with steam condensation supports the theory and conclusions advanced by Thomas.[71] Using 1·0-mm diameter circular wires spaced at nine wire diameters, they measured augmentation ratios in the range of 3–6 on a 1·3-m length test section. The augmentation ratio decreases with increasing heat flux because the wires become more quickly loaded with condensate.

A 1979 US Patent by Notaro[63] describes a new concept—extended surface geometry for enhanced film condensation. It consists of an array of

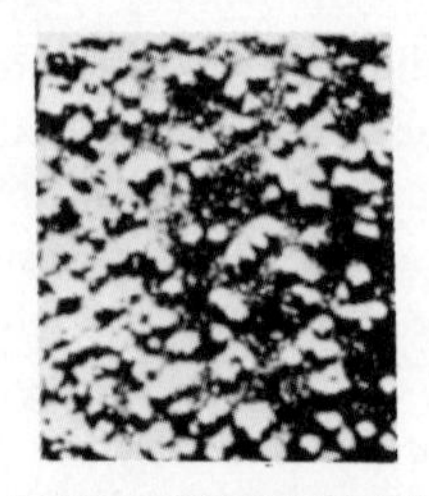

FIG. 7.9. Enhanced condensing surface formed by small-diameter metal particles bonded to the surface.[63] (a) Photo of actual surface; (b) illustration of particles bonded to surface; (c) illustration of thinned condensate film on surface particles and thicker condensate on base surface.

small-diameter metal particles bonded to the tube surface. The particles are 0·25–1·0 mm high covering 20–60 % of the tube surface. Condensation occurs on the particle array and drains along the smooth base surface. High condensation rates occur on the convex surfaces of the particles, due to surface-tension forces which maintain very thin condensate films thereon. The Patent provides performance data for several fluids condensing on vertical tubes. For a given particle height, there is an optimum particle spacing density. A 6·0-m long vertical tube, having a 50 % area density of 0·5-mm diameter particles, yielded a condensation coefficient 17 times that predicted by the Nusselt equation for an equal length smooth tube. Figure 7.9 shows a photograph of the surface, and illustrates the thinned condensate films on the particles.

2.4.3. Forced Condensation Inside Tubes

The augmentation techniques used on the outside of vertical tubes are applicable to condensation inside vertical tubes. Augmentation requirements in horizontal tubes are different, because the force of gravity drains the film transverse to the flow direction. Preferred augmentation techniques for horizontal tubes are yet to be established and commercialized. Those which have been tested include twisted tape inserts, internal fins and integral roughness. Internal fins have yielded the highest augmentation levels.

Internal fins yield high augmentation levels. Vrable *et al.*[77] and Reisbig[78] used R-12 in internally finned tubes. Defining the condensation coefficient in terms of the total surface area, they measured coefficients 20–40 % greater than the smooth-tube value. Accounting for surface area increases of 1·5–2·0, the heat transfer coefficient, based on the nominal tube inside diameter, is 2–3 times the smooth-tube value.

Royal and Bergles[79] condensed steam in four internally finned tubes, three of which had spiralled fins. A later work by Luu and Bergles[80] tested the same tubes with R-113. Their results were generally consistent with the augmentation levels reported for R-12.[77,78] The best internal fin geometry provided 20–40 % higher heat transfer coefficients (total area basis) than the smooth tubes. Preferred internal fin geometries have not been established, although Royal and Bergles provide a correlation of their steam data, which accounts for geometric parameters. Due to the large fluid property differences of steam and R-113, they conclude that different fin geometries may be preferred for water and refrigerants.

Bergles *et al.*[79,80] also tested twisted tapes with steam and R-113. The twisted tape performed distinctly poorer than the internal fins. They gave only 30 % higher coefficients and exhibited pressure drops equal or higher than the internally finned tubes.

Luu[81] found that a repeated-rib roughness increased the R-113 condensing coefficient more than 100 %. For the same augmentation level, roughness may be of greater interest than internal fins, since a roughness requires less tube material.

The 'sand-grain type roughness' of Fenner and Ragi,[9] previously described for single-phase forced convection, was also tested with condensation of R-12. The spaced metal particles provide extended surface at high vapor qualities, and turbulence of the condensate film at lower vapor qualities. The R-12 tests were performed in a 14·5-mm diameter tube having a 50 % area density of $e/D = 0{\cdot}031$ particles. This tube provided enhancement levels (smooth-tube area basis) of 2·4 for low exit qualities (25–60 %) and 4·0 for high exit qualities (60–90 %). The accompanying pressure drops were only 68 % and 105 % larger than the smooth-tube values for the two exit qualities.

2.5. Surface Geometries for Plate Type Exchangers

Modern plate-exchanger geometries normally use some form of

corrugated-plate heat transfer surface. Because these exchangers in liquid service may operate at low Reynolds numbers, e.g. <3000, a surface geometry which provides bulk fluid mixing is required to yield augmentation. Typical plate designs consist of obliquely corrugated plates[83] or 'chevron' corrugated plates.[84] These plates are stacked so that the peaks of adjacent plates are in contact, which allows a moderately high design pressure. Okada *et al.*[83] give basic data on the heat transfer characteristics of simple and obliquely corrugated plates for different wave pitches and heights, and for several wave inclination angles. Plates whose corrugations are oriented transverse to the flow, yield heat transfer coefficients nominally three times greater than a parallel-plate channel of the same plate spacing. For the same wave geometry, the heat transfer coefficient decreases with increasing wave inclination angle. Okada has shown that a 60% decrease occurs as the angle is increased from 15–60°. However, the friction factor falls much faster than the heat transfer coefficient. The friction factor of the 60° wave angle is only 10% as large as that of the 15° plate angle. Therefore, the oblique plates yield much-improved heat transfer/pressure drop ratios. Heat transfer and friction characteristics of a commercially available 'chevron' corrugated plate are given by Marriott.[84]

2.6. Surface Geometries for Parallel Plate-fin Exchangers

The plate-fin exchanger (also known as the 'compact' or 'matrix' exchanger) are similar to the plate-type exchanger. However, in the plate-fin exchanger, at least one of the fluids is normally a gas, and brazed rather than gasketed construction is employed. Extended surface is used to provide additional surface area for heat transfer to gases. Webb presents information on the mechanical design[85] and thermal design[86] of plate-fin exchangers. Augmentation is of interest for the gas-side extended surface. Figure 7.10 shows typical extended surface geometries for gases. The plain rectangular and triangular fin passages offer high heat transfer coefficients due to their small hydraulic diameters. Typical fin densities are 5–8 fins/cm, although as many as 16 fins/cm have been employed for clean gases. The gas flow is normally in the laminar regime. The wavy fin provides increased heat transfer due to secondary flows established by the wavy channel. The offset strip fin, louvered and pin fins all provide augmentation via the repeated growth and wake destruction of laminar boundary layers. The offset strip fin yields heat transfer coefficients 2–3 times higher than a plain fin.

When one of the heat transfer media is a liquid, some variety of parallel-plate flow channel is employed. This may be formed by a stacked-plate

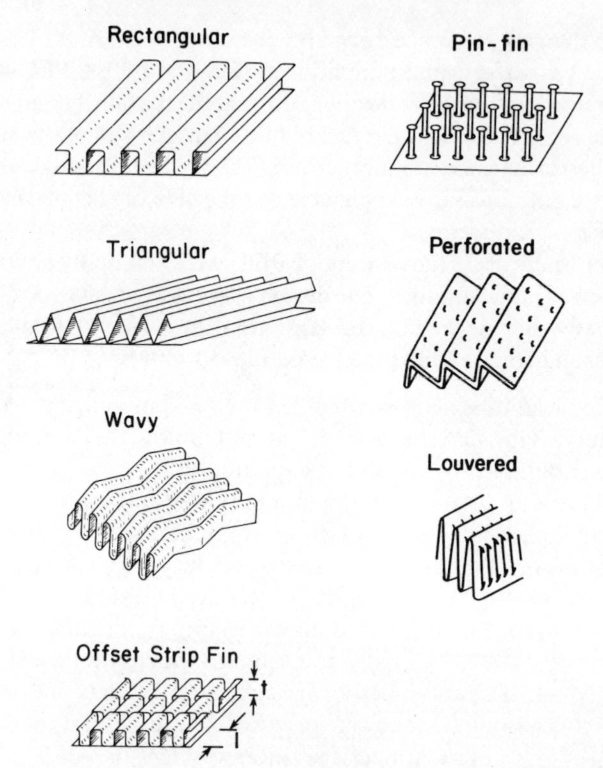

FIG. 7.10. Extended surface geometries used with gases in compact heat exchangers.[86]

construction, or the liquid channel may be an aluminum extrusion. Haberski and Raco[18] describe an innovative gas–liquid design made of an aluminum extrusion, with a variation of the offset strip fin on the gas side. Augmentation is not normally employed on the liquid side. However, Milton's US Patent[25] describes special surface geometries which may be used with boiling and condensing fluids.

3. CRITERIA FOR SELECTION OF AUGMENTED SURFACES

Several factors must be considered to assess the potential for application of augmented surfaces to heat exchangers. First, the designer must establish a performance objective and the necessary design constraints. The objectives are defined such that the augmented surface exchanger is required to do a better job than the reference design for the same constraints. This requires definition of a 'performance evaluation criterion' (PEC). A PEC establishes a quantitative statement to account for the heat transfer improvement, giving due recognition to the pressure loss characteristic of the augmented surface. Performance evaluation criteria established for single-phase flow in tubes are not necessarily applicable to tube-side or shell-side two-phase heat transfer processes.

In order to illustrate the concept of PEC, we will initially limit attention to single-phase flow in tubes. The design objectives are stated relative to a plain tube exchanger having the same total flow rate and entering fluid conditions. There are four possible design objectives:

1. Reduced tube material volume for equal pumping power and heat duty. This may be of little interest unless the size reduction is accompanied by a reduced capital cost.
2. Increased UA for equal pumping power and total length of heat exchanger tubing. A higher UA may be exploited in one of two ways: (a) To obtain increased heat duty. (b) To secure reduced LMTD for fixed heat-duty. Reduced LMTD will effect an improved system thermodynamic efficiency, yielding lower operating costs. A more costly augmented surface will be justified if the savings of operating costs are sufficiently high.
3. Reduced pumping power for equal heat duty and total length of heat exchanger tubing, as in case 2(b) the permissible heat exchanger cost increase is dependent on the pumping power cost savings.

Next, several basic design constraints must be established. These are dependent upon whether the augmentation is being considered for a new design or a retrofit application. Retrofit is defined as a situation where the heat exchanger geometry is fixed (FG). It implies a 1:1 replacement of smooth heat exchanger tubes with augmented tubes. If the reference design has multiple tube-side passes, the retrofit design may have a reduced number of tube-side passes to accommodate the higher friction characteristic of the augmented tubes. In the case of a new design, one may use more tubes per pass to meet the pumping power constraint. This is defined as a 'variable geometry' (VG) case. Table 7.4 (see Section 2) classifies the possible design objectives and constraints. The table contains three PEC which do not require constant heat exchanger flow rates (FG-2, -3 and -4). It may be necessary to operate the augmented exchanger at a reduced flow rate to satisfy the specified constraints. The author's analysis has shown that reduced exchanger flow rate tends to negate the performance offered by the augmented surface. When the flow rate is reduced, the augmented surface exchanger will operate at increased thermal effectiveness. Then it must provide additional *UA* to compensate for the resulting operation at a reduced LMTD. The PEC of Table 7.4 require the augmented exchanger to do a better job than the reference design for equal constraints. For example, the variable geometry (VG) material reduction case requires the augmented tube exchanger to operate at the same flow rate, pumping power and heat duty as the reference exchanger.

In order to calculate the performance improvement, it is necessary to write the objective functions and constraints in terms of algebraic Stanton number and friction factor ratios of the augmented and smooth surfaces. Webb *et al.* give the necessary algebraic relations for internally roughened[7] and internally finned tubes.[8] Bergles *et al.*[87] have defined similar PEC and algebraic relations.

The PEC used for single-phase flows may not be the most appropriate for two-phase exchangers. In two-phase exchangers, pressure drop of the phase-change fluid may cause reduced LMTD. A good example is that of a direct expansion refrigeration machine with tube-side evaporation. Because the compressor suction pressure is fixed, refrigerant pressure drop reduces the LMTD and effects reduced evaporator heat duty. An appropriate PEC should establish the refrigerant flow rate which yields maximum evaporator load. This criterion is employed in the analysis of test results on internally roughened tubes by Withers and Habdas,[45] and by Lord *et al.*[47] Similar arguments may be stated for refrigeration condensers having fixed compressor discharge pressure. Alternatively, the condenser PEC may be based on fixed LMTD with consideration of the increased compressor power required to compensate for the increased pressure drop of the condensing vapor. These considerations may be more important if the phase-change pressure drop is greater for flow in tubes than on the shell side. Significant pressure loss increase has not been reported for use of shell-side enhanced nucleate boiling surfaces. Specific PEC have not been proposed for two-phase situations. However, Royal and Bergles[79] use the single-phase PEC to evaluate several surface geometries for in-tube condensation, and by comparing the tube geometries for a fixed pressure drop, they maintain constant LMTD and compressor power for a refrigeration application.

Normally, one may design an augmented heat exchanger for the same pressure drop as a smooth-tube exchanger if it is possible to allow a modest increase of exchanger flow frontal area. Therefore, constraints must be established for the exchanger geometry. In most cases, the heat exchanger materials are specified for the application. However, the augmentation technique under consideration may be limited to the use of a particular material for its manufacture. Finally, the fouling potential of the heat transfer fluid may further limit use of possible augmentation techniques.

Cost is obviously an important consideration. If the goal is material reduction, augmentation may not be attractive unless a capital cost reduction is realized. If the objective is to reduce operating cost (reduced pumping power, or increased system energy efficiency), more expensive augmented surfaces are justified. The permissible cost increase will depend on economic analysis of the operating cost savings.

4. USE OF AUGMENTED SURFACES IN HEAT EXCHANGERS

Although many augmentation techniques have been evaluated in laboratory studies, the majority have not found commercial application. Further, several techniques are in competition with one another. So, only those which offer the highest performance and lowest cost will find widespread commercial application.

For two-fluid exchangers, the ratio of the thermal resistances is of primary importance in determining whether augmentation will be of benefit. Augmentation should be considered for the fluid stream which has the controlling thermal resistance. If both resistances are approximately equal, augmentation may be considered for both sides of the exchanger.

The present status of special surface geometries may be summarized as follows:

Tube-side enhancement for finned-tube exchangers. Internally finned tubes are commercially available in copper and aluminum. These offer higher performance than wire coil or twisted tape inserts for most applications. Corrugated tubes are not suitable for finned-tube exchangers because the tube cannot be adequately expanded to form a tight fit with the fin. Manufacturing technology has not been developed to make other internal roughness geometries.

Extended surfaces for gases. The use of extended surfaces is well accepted, and has been used for some years. Recent advances in finned tubes include wavy and segmented fins which offer higher performance than smooth fins. For clean gases, the trend has been towards use of staggered tubes and more-closely spaced fins. The compact plate-fin heat exchanger, made of aluminum, is rapidly growing in acceptance. The use of aluminum extrusions permits its use as a gas–liquid exchanger. Use of the segmented fin provides high gas-side performance. New heat exchanger designs using steel materials or duplex material construction expands its range of applications.

Extended surfaces for liquids. Low integral-fin tubing is commonly employed for shell-and-tube exchangers. A wide range of fin spacing is available.

Shell-side condensation. Low integral-fin tubing is well accepted for condensation on horizontal tubes. The trend has been to use more-closely spaced fins for low surface-tension fluids. New technological developments (including saw-tooth or spine fins) are in the early stages of commercial introduction. The recently disclosed concept of distributed small metal particles bonded to the tube surface offers interesting new commercial possibilities. Fluted tubes are commercially employed for condensation on vertical tubes.

Nucleate boiling surfaces. Rapid advances have been made in this area for shell-side boiling. Several varieties of nucleate boiling surfaces are commercially available and they are expected to make integral-fin tubing for boiling applications obsolete.

Doubly augmented tubes. Tubes which have enhancement on both the inner and outer surface are beginning to appear. The availability of such tubes has been limited by manufacturing technology and proprietary interest. One manufacturer presently offers an internally roughened tube (rib-roughness at a helical angle) with external integral fins. Internally finned tubes with porous coatings for nucleate boiling on the outer surface have been commercially employed.

Plate-type exchangers now use corrugated surfaces of special designs to obtain high heat transfer performance. Such designs have been primarily applicable to heat exchange between liquids. However, commercial developments are in progress to provide plate-type exchangers for two-phase service.

5. PROSPECTS FOR THE FUTURE

The development of extended surfaces on the outside of tubes has proceeded on a proprietary basis, and has been successfully commercialized. Empirical efforts have yielded high performance surface geometries. This area is relatively mature, in terms of augmentation technology. However, further work on design methods for optimum surface selection should be continued. Innovative forms of plate-fin heat exchanger geometries will continue to be advanced, especially for gas–liquid or gas–two-phase applications. The use of duplex material construction will expand the range of heat transfer applications.

A high level of R and D activity on high performance boiling and condensing surfaces is in progress. This activity should yield lower-cost surface geometries and a wider range of materials selection.

Most of the previous commercial efforts have focused on the outer surface of the tube. Very little has been done to develop tube-side enhancement, with the exception of internally finned tubes. Internal roughness offers high potential for single-phase forced convection inside tubes. Preferred internal surface geometries for boiling and condensing inside tubes are yet to be established. Finally, doubly augmented tubes (inside and outside enhancement) require extensive development efforts and offer high promise for advanced heat exchanger technology.

Presently, cost effective manufacturing technology is probably the most significant barrier to commercial application of high performance augmentation technology. New manufacturing techniques may be required, especially for tube-side enhancement. Developments in manufacture of doubly augmented tubes are sorely needed, and offer high potential for new heat exchanger technology.

Greater communication is needed between researchers and manufacturing technologists. Researchers have attempted to identify high performance surface geometries without regard to their manufacturability. Future communications which identify acceptable heat transfer performance and permit low manufacturing cost will advance the commercialization of augmentation technology. Additional work is needed on analytical design methods to establish preferred surface geometries, and the optimum geometric parameters of the preferred surfaces. The tools for surface selection must be moved from the researcher and put in the hands of heat exchanger designers.

REFERENCES

1. Bergles, A. E. and Webb, R. L. Energy conservation via heat transfer enhancement, *Energy*, 1979, **4**, 193–200.
2. Bergles, A. E. and Webb, R. L. Bibliography on augmentation of convection heat and mass transfer, Report No. HTL-19, Engineering Research Institute, Iowa State Univ., Ames, Iowa, May 1979.

3. KUMAR, P. and JUDD, R. L. Heat transfer with coiled wire turbulence promoters, *Can. J. Chem. Eng.*, 1970, **8,** 378–83.
4. LOPINA, R. F. and BERGLES, A. E. Heat transfer in tape generated swirl flow of single-phase water, *J. Heat Trans.*, 1969, **91,** 434–42.
5. HONG, S. W. and BERGLES, A. E. Augmentation of laminar flow heat transfer in tubes by means of twisted-tape, inserts in, *J. Heat Trans.*, 1976, **98,** 251–56.
6. WEBB, R. L. Toward a common understanding of the performance and selection of roughness for forced convection, in: *A Fest-schrift for E. R. G. Eckert*, J. P. Hartnett, T. F. Irvine, Jr., E. Pfender and E. M. Sparrow (eds.), Hemisphere, Washington D.C., 1979.
7. WEBB, R. L. and SCOTT, M. J. A parametric analysis of the performance of internally finned tubes for heat exchanger application, *J. Heat Trans.*, 1980, **102,** 38–43.
8. WEBB, R. L. and ECKERT, E. R. G. Application of rough surfaces to heat exchanger design, *Int. J. Heat–Mass Trans.*, 1972, **15,** 1647–58.
9. FENNER, G. W. and RAGI, E. Enhanced tube inner surface heat transfer device and method, US Patent 4,154,293, May 15 1979.
10. CARNAVOS, T. C. Heat transfer performance of internally finned tubes in turbulent flow, in: *Advances in Enhanced Heat Transfer*, J. M. Chenoweth, J. Kaellis, J. Michel and S. Shenkman (eds.), 1979, ASME, New York, 61–68.
11. GEE, D. L. and WEBB, R. L. Forced convection heat transfer in helically rib-roughened tubes, 1980, *Int. J. Heat–Mass Trans.* (in press).
12. MARNER, W. J. and BERGLES, A. E. Augmentation of tube-side laminar flow heat transfer by means of twisted tape inserts, static mixers inserts, and internally finned tubes, *Heat Transfer*, 1978, **2,** 583–88.
13. WITHERS, J. G. Jr. and REIGER, K. K. Heat transfer tube having multiple internal ridges, US Patent 3,847,212, Nov. 12, 1974.
14. ANON. Boiling and steam separation, in: *Steam—Its Generation and Use*, 1978, Babcock and Wilcox Pub., New York, 5.
15. MCLAIN, C. D. Process for preparing heat exchanger tube, US Patent 3,906,605, Sept. 23, 1975.
16. WEIERMAN, C., TABOREK, J. and MARNER, W. J. Comparison of the performance of inline and staggered banks of tubes with segmented fins, AICHE Paper No. 6, 15th Nat. Heat Trans. Conf., San Francisco, Aug. 10, 1975.
17. BRAUER, A. Compact heat exchangers, *Chem. Process Eng.*, 1964, 451–60.
18. HABERSKI, R. J. and RACO, R. J. Engineering analysis and development of an advanced technology, low cost, dry-looking, tower heat-transfer surface, ERDA Report COO-2774-1, Curtiss-Wright Corp., Mar. 1, 1976.
19. WEBB, R. L. Air-side heat transfer in finned-tube heat exchangers, *Heat Trans. Eng.*, 1980, **1**(3), 33–40.
20. GROEHN, H. G. and SCHOLZ, F. Heat transfer and pressure drop of inline banks of tubes with artificial roughness, in: *Heat and Mass Transfer Source Book: 5th All Union Conference, Minsk, 1976*, M. A. Styrikovich, A. Zukauskis, J. P. Hartnett and T. F. Irvine, Jr. (eds.), Hemisphere, Washington D.C.
21. WHITE, L. and WILKIE, D. The heat transfer and pressure loss characteristics of some multi-start ribbed surfaces, in: *Augmentation of Convective Heat and Mass Transfer*, 1970, ASME, New York, 55–62.

22. Williams, F., Pirie, M. A. M. and Warburton, C. Heat transfer from surfaces roughened by ribs, in: *Augmentation of Convective Heat and Mass Transfer*, 1970, ASME, New York, 55–62.
23. Webb, R. C., Eckert, E. R. G. and Goldstein, R. V. Generalized heat transfer and friction correlations for tubes with repeated-rib roughness, *Int. J. Heat–Mass Trans.* 1972, **15,** 180–84.
24. Czikk, A. M., Gottzman, C. F., Ragi, E. G., Withers, J. G. and Habdas, E. P. Performance of advanced heat transfer tubes in refrigerant flooded liquid coolers, *Trans. ASHRAE*, 1970, **1**(1), 96–109.
25. Milton, R. M. Heat exchange system, US Patent 3,523,577, Aug. 11, 1970.
26. Gottzman, C. F., Wulf, J. B. and O'Neill, P. S. Theory and application of high performance boiling surfaces to components of absorption cycle air conditioners, 1971, Proc. Conf. on Natural Gas Research and Technology, Session I, Paper 3.
27. Milton, R. M. and Gottzman, C. F. High efficiency hydrocarbon reboilers and condensers, *Chem. Eng. Prog.*, 1972, **68**(9), 55–61.
28. Dahl, M. M. and Erb, L. D. Liquid heat exchanger interface method, US Patent 3,990,862, Nov. 9, 1976.
29. Janowski, K. R. and Shum, M. S. Heat transfer surface, US Patent 4,129,181, Dec. 12, 1978.
30. Webb, R. L. Heat transfer surface which promotes nucleate ebullition, US Patent 3,521,708, Oct. 10, 1972.
31. Fujii, M., Nishiyama, E. and Yamanaka, G. Nucleate pool boiling heat transfer from a microporous heating surface, in: *Advances in Enhanced Heat Transfer*, 1979, J. M. Chenoweth, J. Kaellis, J. Michel and S. Shenkman (eds.), ASME, New York.
32. Arai, N., Fukushima, T., Arai, A., Nakajima, T., Fujii, K. and Nakayama, Y. Heat transfer tubes enhancing boiling and condensation in heat exchangers of a refrigerating machine, *Trans. ASHRAE*, 1977, **83**(2), 58–70.
33. Torii, T., Hirasawa, S., Kuwahara, H. and Yanagida, T. The use of heat exchangers with THERMOEXCEL's tubing in ocean thermal energy power plants, ASME Paper 78-WA/HT-65, 1978.
34. Anon. GEWA-T-tubes: high performance tubes for flooded evaporators, Brochure SAW-15e-06.78, Wieland-Werke AG, Metabuerke, Ulm, West Germany, 1978.
35. Kun, L. C. and Czikk, A. M. Surface for boiling liquids, US Patent 3,454,081, July 8, 1969.
36. Alexander, L. G. and Hoffman, H. W. Performance characteristics of corrugated tubes for vertical tube evaporators, ASME Paper 71-HT-30, 1971.
37. Rifert, V. G., Butuzov, A. I. and Belick, D. N. Heat transfer in vapor generation in a falling film inside a vertical tube with a finely-finned surface, *Heat Transfer—Soviet Research*, 1975, **7**(2), 22–25.
38. Sephton, H. H. Vertical tube evaporation with double fluted tube and interface enhancement, ASME Paper 75-HT-43, 1975.
39. Thomas, D. G. and Young, G. Thin film evaporation enhancement by finned surfaces, *IEC Process Des. Dev.*, 1970, **9,** 317–23.
40. Conti, R. J. Heat transfer enhancement in horizontal ammonia film evaporators, *Proc. 6th OTEC Conf.*, Paper 5D-2, Washington D.C., June 19–22, 1979.

41. EDWARDS, D. K., GIER, K. D., AYYASWAMY, P. S. and CATTON, I. Evaporation and condensation in circumferential grooves on horizontal tubes, ASME Paper 73-HT-25, 1973.
42. SCHULTZ, V. N., EDWARDS, D. K. and CATTON, I. Experimental determination of evaporative heat transfer coefficients on horizontal, threaded tubes, AIChE Paper No. 17, 16th Nat. Heat Trans. Conf., St. Louis, 1976.
43. O'NEILL, P. S., RAGI, E. G. and JACOBS, M. L. Effective use of high flux tubing in two-phase heat transfer, AIChE Paper presented at Session 83, 86th National Meeting, Houston, April 1979.
44. MURPHY, R. W. and BERGLES, A. E. Subcooled flow boiling of fluorocarbons: hysterisis and dissolved gas effects on heat transfer, Proc. 1972 Heat Trans. and Fluid Mech. Inst., 1972, Stanford Univ. Press, Stanford, C., 400–16.
45. WITHERS, J. G. and HABDAS, E. P. Heat transfer characteristics of helical corrugated tubes for in-tube boiling of refrigerant-12, *AIChE Symp. Ser*, 1974, **70**(138), 98–106.
46. ITO, M., KIMURA, H. and SENSHU, T. Development of high efficiency air-cooled heat exchangers, *Hitachi Review*, 1977, **26**(10), 323–6.
47. LORD, R. G., BUSSJAGER, R. C. and GEARY, D. F. High performance heat exchanger, US Patent 4,118,944, Oct. 10, 1978.
48. KUBANEK, G. R. and MILLETTI, D. C. Evaporative heat transfer and pressure drop performance of internally-finned tubes with refrigerant-22, ASME Paper 77-WA/HT-25, 1977.
49. LAVIN, J. G. and YOUNG, E. H. Heat transfer to evaporating refrigerants in two-phase flow, *AIChE J.*, 1965, **11,** 1124–32.
50. SCHLÜNDER, E. U. and CHWALA, J. M. Local heat transfer and pressure drop for refrigerants evaporating in horizontal internally finned tubes, *Int. Cong. Refrig.*, 1967, 601–12.
51. BRYAN, W. L. and SEIGEL, L. G. Heat transfer coefficients in horizontal tube evaporators, *Refrigeration Eng.*, May, 1955, 36–45 and 126.
52. LARSON, R. L., QUAINT, G. W. and BRYAN, W. L. Effects of turbulence promoters in refrigerant evaporator coils, *J. ASRE*, Dec. 1949, 1193–95.
53. PEARSON, J. F. and YOUNG, E. H. Simulated performance of refrigerant-22 boiling inside tubes in a four pass shell-and-tube heat exchanger, *AIChE Symp. Ser.*, 1970, **66**(102), 164–73.
54. GAMBILL, W. R. and GREEN, N. D. Boiling burnout with water in vortex flow, *Chem. Eng. Prog.*, 1958, **54**(10), 58–76.
55. MAYINGER, F., SCHAD, O. and WEISS, E. Investigations into the critical heat flux in boiling, Maschinenfabrik Augsburg-Nuernbuerg Final Report No. 09.03.01, May 1966.
56. BERGLES, A. E., FULLER, W. O. and HYNEK, S. W. Dispersed flow film boiling of nitrogen with swirl flow, *Int. J. Heat–Mass Trans.*, 1971, **14,** 1345–54.
57. BLATT, T. A. and ADT, R. A. JR. The effects of twisted tape swirl generators on the heat transfer rate and pressure drop of boiling freon-11 and water, ASME Paper 63-WA-42, 1963.
58. NOORUDDIN, A. F. and MURTI, P. S. Heat transfer to gas–liquid mixtures in a vertical tube fitted with twisted tapes, *Int. J. Heat–Mass Trans.*, 1973, **16,** 1655–57.
59. TANASAWA, I. Dropwise condensation—the way to practical applications, Proc. 6th Int. Heat Trans. Conf., 1978, Hemisphere, Washington D.C.

60. Iltscheff, S. Concerning tests for achieving drop condensation with fluorinated refrigerants, *Kaltetechnick-Klimatisierung*, 1971, **23,** 237–41.
61. Beatty, K. O. and Katz, D. L. Condensation of vapors on outside of finned tubes, *Chem. Eng. Prog.*, 1948, **44**(1), 55–70.
62. Taborek, J. Design methods for heat transfer equipment, Ch. 3 in: *Heat Exchangers: Design and Theory Sourcebook*, N. Afgan and E. U. Schlünder (eds.), 1974, McGraw-Hill, New York, 69.
63. Notaro, F. Enhanced condensation heat transfer device and method, US Patent 4,154,294, May 15, 1979.
64. Torii, T., Hirasawa, S., Kuwahara, H., Yanagida, T. and Fujie, F. The use of heat exchangers with THERMOEXCEL's tubing in ocean thermal energy power plants, ASME Paper 78-WA/HT-65, 1978.
65. Webb, R. L. and Gee, D. L. Analytical predictions for a new concept spine-fin surface geometry, 1979, *ASHRAE Trans.*, **85**(2), 274–83.
66. Withers, J. G. and Young, E. H. Steam condensing on vertical rows of horizontal corrugated and plain tubes, *IEC Process Des. Dev.*, 1971, **10**(1), 19–30.
67. Marto, P. J., Reilly, D. J. and Fenner, J. H. An experimental comparison of enhanced heat transfer condenser tubing, in: *Advances in Enhanced Heat Transfer*, J. M. Chenoweth, J. Kaellis, J. Michel and S. Shenkman (eds.), 1979, ASME, 1–10.
68. Thomas, A., Lorenz, J. J., Hillis, D. A., Young, D. T. and Sather, N. F. Performance tests of the 1 Mwt shell-and-tube exchangers for OTEC, Paper 3C, Proc. 6th OTEC Conf., June 19–22, 1979.
69. Gregorig, R. Hautkondensation an feingwellten oberflächen bei Beruksichtigung der Oberflachenspannungen, *Zeitschrift fur Angewandte Mathematik und Physik*, 1954, **5,** 36–49.
70. Webb, R. L. A generalized procedure for the design and optimization of fluted gregorig condensing surfaces, *J. Heat Trans.*, 1979, **101,** 335–39.
71. Thomas, D. G. Enhancement of film condensation rate on vertical tubes by longitudinal fins, *J. AIChE*, July 1968, 644–49.
72. Carnavos, T. C. Ch. 17 in: *Heat Exchangers: Design and Theory Sourcebook*, N. Afgan and E. U. Schlünder (eds.), 1974, McGraw-Hill, New York, 441–90.
73. Combs, S. K. and Murphy, R. W. Experimental studies of OTEC heat transfer condensation of ammonia on vertical fluted tubes, Proc. 5th OTEC Conf., Feb. 20–22 1978, Vol. 3 Sect. 6, 111–22, Miami Beach, Fa.
74. Thomas, D. G. Enhancement of film condensation rates on vertical tubes by vertical wires, *Ind. and Eng. Chemistry-Fundamentals*, Feb. 1967, 97–102.
75. Butizov, A. I., Rifert, V. G. and Leont'yev, G. G. Heat transfer in steam condensation on wire-finned vertical surfaces, *Heat Transfer Soviet Research*, 1975, **7**(5), 116–20.
76. Rifert, V. G. and Leont'yev, G. G. An analysis of heat transfer with steam condensing on a vertical surface with wires to promote heat transfer, *Teploenergetika*, 1976, **23**(4) 74–80.
77. Vrable, D. A., Yang, W. J. and Clark, J. A. Condensation of refrigerant-12 inside horizontal tubes with internal axial fins, *Heat Transfer* 1974, 5th Int. Heat Trans. Conf., Vol. 3, 250–54, Japan Soc. Mech. Engrs., Tokyo, 1974.

78. Reisbig, R. L. Condensing heat transfer augmentation inside splined tubes, Paper 74-HT-7, AIAA/ASME Thermophysics Conf., July 1974, Boston.
79. Royal, J. H. and Bergles, A. E. Augmentation of horizontal in-tube condensation by means of twisted tape inserts and internally finned tubes, *J. Heat Trans.*, 1978, **100,** 17–24.
80. Luu, M. and Bergles, A. E. Experimental study of the augmentation of the in-tube condensation of R-113, *ASHRAE Trans.*, 1979, **85**(2), 132–46.
81. Luu, M., Augmentation of in-tube condensation of R-113, PhD Thesis, Dept. of Mech. Eng., Iowa State University, 1979.
82. Johnson, B. M., Jansen, G. and Owzarski, P. C. Enhanced evaporating film heat transfer from corrugated surfaces, ASME Paper 71-HT-33, 1971.
83. Okada, K., Ono, M., Tomimura, T., Okuma, T., Konno, H. and Ohtani, S. Design and heat transfer characteristics of new plate heat exchanger, *Heat Transfer—Japanese Research*, 1972, **1**(1), 90–95.
84. Marriott, J. Performance of a mixed plate heat exchanger, *Chem. Eng. Progress*, Feb. 1977, 73–78.
85. Webb, R. L. Compact heat exchangers—mechanical design and construction, Ch. 3, Part 4, *Heat Transfer Design and Data Book*, ed., E. U. Schlünder, 1980, Hemisphere, Washington D.C.
86. Webb, R. L. Matrix heat exchangers—thermal and hydraulic design, Ch. 9, Part 3, *Heat Transfer Design and Data Book*, ed., E. U. Schlünder, 1980, Hemisphere, Washington D.C.
87. Bergles, A. E., Blumenkrantz, A. R. and Taborek, J. Performance evaluation criteria for enhanced heat transfer surfaces, *Heat Transfer* 1974, Vol. 2, 234–28, Japan Soc. Mech. Engrs., Tokyo, 1974.
88. Nishikawa, K., Yoshida, H. and Ohnu, M. Improvement in heat transfer performance at high heat fluxes with internally grooved boiler tubes, *Memoirs Faculty Eng., Kyushu Univ.*, 1975, **35**(2), 37–49.
89. Gupta, R. K. and Rao, M. R. Heat transfer and friction characteristics of newtonian and power-law type of non-newtonian fluids in smooth and spirally corrugated tubes, in: *Advances in Enhanced Heat Transfer*, J. M. Chenoweth, J. Kaellis, J. Michel and S. Shenkman (eds.), 1979, ASME, New York, 103–13.
90. Czikk, A. M. and O'Neill, P. S. Correlation of nucleate boiling from porous metal films, in: *Advances in Enhanced Heat Transfer*, J. M. Chenoweth, J. Kaellis, J. Michel and S. Shenkman (eds.), 1979, ASME, New York, 53–60.
91. Nakayama, W., Daikoku, T., Kuwahara, H. and Nakajima, T. Dynamic model of enhanced boiling heat transfer and porous surfaces, in: *Advances in Enhanced Heat Transfer*, J. M. Chenoweth, J. Kaellis, J. Michel and S. Shenkman (eds.), 1979, ASME, New York, 31–44.
92. Fricke, H. D. and Czikk, A. M. Enhanced sprayed bundle evaporator performance studies, in: *Advances in Enhanced Heat Transfer*, J. M. Chenoweth, J. Kaellis, J. Michel and S. Shenkman (eds.), 1979, ASME, New York, 23–30.
93. Panchal, C. B. and Bell, K. J. Analysis of Nusselt-type condensation on a vertical fluted surface, *Condensation Heat Transfer*, P. J. Marto and P. G. Kroeger (eds.), 1979, ASME, New York, 45–54.

Chapter 8

HEAT PUMPS

P. Freund†
Building Research Establishment, Watford, UK

SUMMARY

One of the techniques which offers great potential for energy conservation is the heat pump and this provides a major application for heat exchanger technology. The heat pump extracts energy from a source at low temperature, too low for it to be useful for heating purposes directly, and upgrades it to a higher temperature. Typically, two to five times as much energy is provided for heating as is consumed in driving the heat pump. This is what makes the device attractive as a means of reducing energy consumption.

The uses of heat pumps range from room heating, through space conditioning to process heating and industrial drying. In this chapter the principles of heat pump operation are illustrated with examples based on the commonly used reversed-Rankine cycle. Applications and sources of low temperature energy are discussed, practical details and likely performance are described. A brief outline of developments now underway is included.

1. INTRODUCTION

In modern society great importance is attached to the notion of 'energy' although it is essentially an abstract concept. Energy provides goods and services which are the concrete representation of the user's needs and, in this respect, the principle of pumping heat has been applied for many years to obtain a particular service, viz. refrigeration. It has long been recognised that the same technique makes possible the pumping of heat from a low temperature to a higher temperature so that it becomes useful for heating

† Present address: BP Research Centre, Chertsey Road, Sunbury-on-Thames, Middlesex, UK.

purposes. This offers a significant potential for reducing energy consumption for heating, but it is only recently that heat pumps have been used in any numbers specifically to provide heating rather than cooling.

The second law of thermodynamics shows that work must be done in moving heat from a cold source to a warmer sink[1] (the Clausius statement). The heat pump obtains energy from the low-temperature source which is virtually free and produces, perhaps, two to three times as much heat as the high-grade energy consumed in driving it—and so the attraction in terms of energy saving is clear.

The heat pump is a more complicated device than conventional heating appliances and its first cost is correspondingly higher. Thus it can only compete if the cost of energy is also high. The uses to which it may be put range from domestic space conditioning and service water heating through heat reutilisation in refrigeration plant to purpose-designed units for industrial process heating. In the following sections the principles of the heat pump are outlined and some of the practical aspects are examined; a few applications are illustrated and, finally, a brief review of current developments shows what may be expected in the future.

2. PRINCIPLES OF THE HEAT PUMP

The concept of a heat pump was first propounded by Lord Kelvin and an idealised cycle can be used to illustrate the thermodynamics involved. In Fig. 8.1 the Carnot cycle for a heat pump is shown—from this it can be seen that the heat given out in the process is $Q_2 = T_2(S_2 - S_1)$ and the work done (represented by the enclosed area) is $W = (S_2 - S_1)(T_2 - T_1)$.

The effectiveness of the machine is measured by the coefficient of performance for heating COP_h, which is $COP_h = Q_2/W$. For the Carnot cycle the $COP_h = T_2/(T_2 - T_1)$ but in practice the 'work done' includes all the high-grade energy (such as electricity) consumed in operating the heat pump. There are many thermodynamic cycles that could be used for a heat pump and, as a measure of system efficiency, the coefficient of performance proves to be a handy description of individual machines. Another parameter, which is useful when comparing different types of heat pump and other heating appliances, is the primary energy ratio (PER) which is

$$\text{PER} = \frac{\text{heat given out}}{\text{primary energy consumed}}$$

The COP_h is a function of the working temperature and, more importantly, of the temperature difference $(T_2 - T_1)$ as shown in Fig. 8.2.

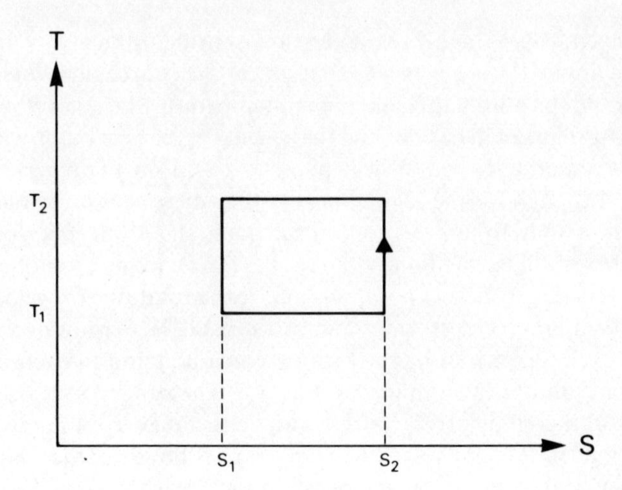

FIG. 8.1. The Carnot cycle for a heat pump on a temperature (T)–entropy (S) diagram.

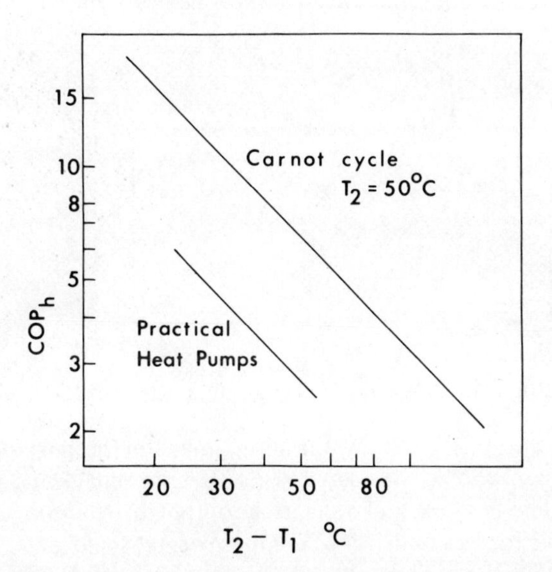

FIG. 8.2. The variation of coefficient of performance (COP_h) with source–sink temperature difference ($T_2 - T_1$) for the Carnot cycle and typical practical heat pumps.

This dependence is highly relevant to any consideration of how heat pumps may be applied, since it means that unless the source of low-temperature energy and the output (or sink) of high-temperature heat are close together, the performance will be poor and the benefits of the heat pump will be slight.

The commonest type of heat pump is based on the reversed-Rankine cycle—although for the sake of brevity the adjective is often omitted. The discussion that follows is concentrated on this cycle but some of the alternatives will be mentioned in passing. The Rankine cycle offers efficient heat transfer from evaporation and condensation processes but its theoretical limit of performance falls short of the Carnot cycle ideal.

The cycle takes four distinct stages corresponding to each of the four principal components outlined in Fig. 8.3. The system is a closed one and the working fluid (the refrigerant) circulates continuously during operation. In the first stage, the refrigerant is boiled in the evaporator by intake of heat from the low-temperature source; in the second stage vapour is drawn from the evaporator into the compressor where it is raised to such a pressure that (at the third stage) it can condense and give out heat at a temperature high enough to be useful; in the final stage, the high-pressure liquid passes through an expansion device (e.g. an orifice) to lower its pressure, thus completing the cycle back to the evaporator.

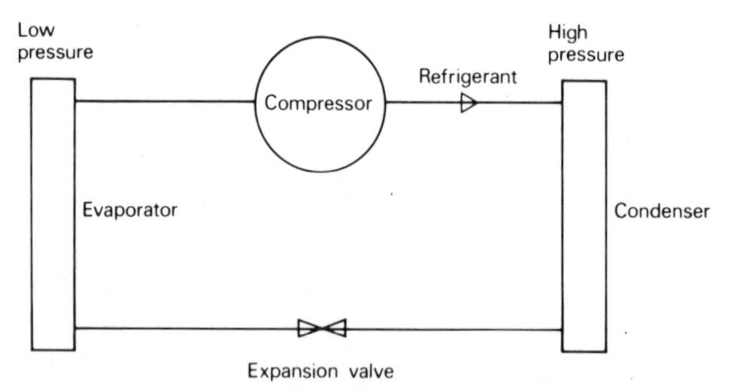

FIG. 8.3. A schematic diagram of a heat pump.

The details are open to many variations, but for the purposes of the example, a system will be considered which uses the outside air as source, the air in a building as sink, and water as the output distribution medium. It also has electric motors to drive the compressor, fans and water pump. If the source is at 0 °C and the output at 50 °C, the Carnot cycle $COP_h = 6{\cdot}46$. In practice the COP_h may be about 40 % of this value (i.e. about 2·5) and it is worth considering what the reason is for this discrepancy. As the building is at about 20 °C the Carnot COP_h for the task is 14·7, but the necessity of using heat distribution systems of finite size requires an output temperature well above 20 °C with consequent reduction in COP_h. The 'real' Rankine cycle differs from the ideal Carnot cycle in that non-isothermal and non-isentropic processes are involved with a real fluid and about 30 % of the difference between the two may be ascribed to this cause. The motors for driving fans and pumps represent a 20 % increase in power consumption over the ideal cycle, and the inefficiency of the motor–compressor unit accounts for a further 20 %. Temperature differences at the condenser and the evaporator, necessary to transfer heat to and from the refrigerant, are the cause of another 15 %. The change in temperature of the source medium accounts for a further 15 %. Because the output temperature is set at 50 °C for this example, the temperature rise of the output medium does not occur in this list, but represents a further 20 % compared with an ideal case with no such rise.

With all these degradations in performance it might be asked whether the Rankine cycle is the best one to use—in practice the alternatives have similar drawbacks and, just as importantly, the technology exists and is widely used in the refrigeration and air conditioning industry. Some alternatives are the Brayton cycle, the absorption cycle and the Peltier effect. The Brayton cycle is operated entirely in the gas phase so that the heat transfer processes are non-isothermal, the equipment is bulkier and the performance poorer than the Rankine cycle. The absorption cycle is a variant of the (reversed) Rankine cycle in which fluid movement is achieved by absorption in a carrier liquid and subsequent regeneration. The Peltier effect is found between junctions of dissimilar conducting materials when an electrical current is passed through them; the heat pumping action is related to the voltage generated at such thermocouple junctions when a temperature difference is maintained across them. The effectiveness of these heat pumps is lower than that of the Rankine cycle but each has particular applications where, because of its size or the type of high-grade energy, it is a practicable option.

The particular features of the Rankine cycle which have led heat pumps to follow refrigeration technology are the mass production of electric motor–compressor units and refrigerants. The requirements for heat exchangers are related to those of the refrigeration industry but such specialised components would be needed whatever the cycle chosen. The refrigerants used are chlorofluorocarbons (e.g. CCl_2F_2 and $CHClF_2$), since these offer good performance whilst being safe and non-toxic, require only moderate working pressures and do not need special materials. The demand for these in refrigeration is such that they are produced relatively cheaply in large quantities. Compounds such as ammonia and sulphur dioxide have disappeared from use except in some large refrigeration plant where the associated problems can be adequately handled.

Continuing with the example given above, a heat pump using refrigerant R-12 (CCl_2F_2) would have the fluid in the evaporator at a pressure of about 0·24 MPa (i.e. boiling at −7 °C). The outside air at 0 °C would be cooled to about −4 °C in boiling the refrigerant and the vapour entering the compressor would be at a temperature of about −3 °C (i.e. slightly superheated). The compressor output at 1·3 MPa with a vapour temperature of 62 °C would be fed to the condenser. The liquid condenses at 52 °C and leaves slightly subcooled (45 °C) to pass through the pressure reducer and return to the evaporator. This simplified description has ignored the small pressure drops that occur throughout the system. Although some of the heat input/output arises from sensible heating/cooling of the fluid, most of it is due to the transfer of latent heat in the phase changes.

Having examined the heat pump cycle, it is now appropriate to discuss the types of source and sink available. It is important to choose pairs separated by small temperature differences but it is also necessary to allow for the thermal capacity of each (this affects the temperature difference across the heat exchangers) and the absolute temperature of the source (this determines the density of the refrigerant). As the compressor circulates a roughly constant volume of vapour, the heat output is a strong function of density and hence source temperature. Although this does not affect the COP_h directly it is likely to influence seasonal performance (in a space-heating device) and plant capacity and cost. The heat pump only performs tasks which conventional heating appliances could also do, but the range of possibilities for combining types of source, sink and application provides great scope for innovation.

The selection of a source will depend upon the task to be performed and the geographical position, cost, specific heat and density, as well as temperature. Some typical combinations are shown in Table 8.1.

In mild climates air is the source most likely to be used for space and water heating because of its availability. In the USA a major market for heat pumps has grown up using air as the source because the same hardware can, with minor modifications, provide heating in winter and cooling in summer.[2] These dual-purpose heat pumps will be discussed further below.

The ground provides a steady temperature at depths of 2 m or more so that it may be used as a heat source for buildings which have enough land area available.[3] Its storage capacity may be exploited for limited periods if sufficient heat is gained from the environment during the heating season to prevent the ground temperature falling too low but, in severe climates, systems which make use of the ground for inter-seasonal storage have been employed—these require considerably larger volumes.[4] Artificial storage in the form of water or ice has been used in certain cases[5] but the capital cost is very high if the utilisation is limited to one cycle of the store per year.

TABLE 8.1

Task	*Requirement*	*Typical source*	
Space and water heating	40–70 °C	Ground	0–10 °C
		Air	−10–15 °C
		Waste heat	10–30 °C
Process heating	50–100 °C	Waste heat	20–70 °C
Dehumidification	30–50 °C	Waste heat	20–40 °C

Ground water is an attractive source to the user but water supply considerations would probably prevent large-scale application. Waste heat is a potentially fruitful source for industrial heat pumps[6] and the introduction of domestic mechanical ventilation would provide a route for applications in the home also.[7] Although the heat pump could be used to reduce the temperature of the waste to that of the environment, this will not necessarily be the best approach. Bearing in mind that the temperature of evaporation controls both the heat putput and the COP_h it is possible to see that the optimum heat pump is one which makes use of only a fraction of the waste heat, thus leaving the residual at above ambient temperature. In commercial premises, such as supermarkets and cold stores, water heating may be obtained as a by-product of refrigeration but in domestic systems the demand for these two services is not sufficiently well matched. Packaged refrigeration equipment usually rejects heat into its immediately surrounding area so providing some useful heat directly.

For domestic heat pumps, the principal source would seem to be the outside air, if they are to be applied on a wide scale.[8] Its low thermal capacity means that large volumes must be moved, with consequent noise problems, and the energy consumed in doing this is an important part of the total consumption. Although the moisture present in the air improves the performance by providing latent heat of condensation (thereby raising the evaporating temperature compared with sensible cooling of the air only), it also presents problems with frost formation on the evaporator. This occurs if the air is within a few degrees of 0 °C and serves to impede the flow and reduce performance. Automatic means of defrosting must be built in but the effect on annual consumption is small.[9]

The reduction in evaporating temperature and heat output on the coldest days means that a method of supplementing the heat pump may be required, for instance additional heaters[10] or an alternative heating system.[11] The latter method allows the heat pump to operate as a base load system giving high efficiency but requires extra capital expenditure. For this reason in mild climates a compromise is reached with supplementary heat providing, perhaps, 10 % of the annual heat supply, which allows the heat pump to be sized for about 80 % of the design heat loss with a consequent saving in capital cost.

The method of supplementing a space heating heat pump is an important feature of the system—directly acting electric heating imposes undue peak loads on the supply network so that other methods will be required before widespread use is acceptable. This must be allowed for in designing the system since without automatic provision of a supplement the users would probably operate electric heaters themselves and arrive at the same end result of a poor load factor. The reaction of the supply authorities to this might be to impose demand-related tariffs for heat pumps (as those in some countries already do).

Supplementary heat may also be used to provide rapid response, for instance on start-up. From the energy point of view, the system is likely to be more efficient if the change in demand can be anticipated and met by the heat pump, and for this reason night set-back of the thermostat may not be as beneficial for heat pumps as for conventional appliances.[12]

The output temperature is limited by the refrigerant and lubricating oil stability and by the laws of thermodynamics. For space heating, 60 °C is the normal upper limit so that extended-surface radiation or warm air distribution is required. Higher output temperatures are now becoming practicable for industrial processes[13] where high source temperatures are available, and heat pumps capable of working at over 100 °C should expand the range of processes into which they can be introduced.

3. PRACTICAL ASPECTS OF HEAT PUMP DESIGN

The details of one particular type will be discussed as an illustration of the practical features of heat pumps. The external air-source water-heating unit is now produced in a number of European countries as a domestic central heating appliance.[14] The machine may be assembled as a single package or as two units (one indoors, one outdoors) connected by on-site pipework. The former type could be expected to be more reliable in the long term, as it offers the prospect of all-brazed connections and factory charging and testing.

The evaporator is mounted outside the building, to provide easy access to the air (Fig. 8.4), or in the roof space to benefit from solar gain[15]—the penalty to be paid with this approach is that this may encourage increased

FIG. 8.4. The evaporator of an external air-source heat pump. (Reproduced by permission of the Building Research Establishment.)

heat transfer from the conditioned space so that some of the 'pumped' heat is recycled through the machine. The net result is to replace some energy that would have come from the 'free' heat source and this has a deleterious effect on the performance.

The heat transfer coefficient on the refrigerant side of the heat exchanger is considerably higher than on the air side, which leads to a requirement for extended surfaces (with a ratio on the two sides of up to 10:1). Compact heat exchangers using aluminium-finned copper tube are common at present.[16] Various detail improvements are possible to increase the effectiveness of the heat exchangers[17] and, in heat pumps designed principally for cooling, a fin spacing of only 2 mm is often used. In mild humid climates, evaporators with such closely spaced fins suffer badly from frosting during heating operation. A simple solution is an increase in fin spacing, thereby reducing the frequency of defrosting, but the consequent heat exchanger is more bulky and less easily installed close to a dwelling.

The condenser, for rejecting heat to water, is often a shell-and-tube or tube-in-tube design. Extended surfaces may not be needed or may be used on the refrigerant side. Additional heat exchangers in the heat pump circuit provide subcooling of the liquid refrigerant, by superheating the vapour entering the compressor, and desuperheating of the compressor output, for instance to provide domestic hot water at a temperature above that of the condenser.[18] The use of subcooling to provide heating directly would seem to be a more straightforward approach for heating-only machines.

Compressors are commonly of the reciprocating type, but rotary and sliding vane types are available for the smaller sizes; in large plant, centrifugal or screw compressors may be used within limited ranges of compression ratios. The drive is most conveniently provided by an electric motor but, in many cases, higher PER is obtained if combustion engine drives can be used with on-site reclamation of the waste heat.[19] So far this has been practicable only in large plant subject to regular maintenance but it is hoped that reliable packaged units for domestic use will be available eventually. The electric motor–compressor units adapted from refrigeration technology[20] are produced in welded (hermetic) or flanged and gasketted (semi-hermetic) casings. The former method is cheaper and more likely to be the practice for mass production; if a fault occurs the whole motor–compressor is replaced rather than repaired, as might be done with the semi-hermetic unit.

The pressure reducer in the smallest units is a capillary tube but, at larger sizes, a variable orifice controlled by feedback from the vapour leaving the evaporator. This is a justifiable improvement since it permits better control over a wider range. These thermostatic expansion valves are normally operated by vapour pressure in a bulb attached to the evaporator outlet, but recent developments involve a method of detecting the presence of liquid in the outlet line with electrical feedback.

Control of the heat pump is by on–off instruction in smaller units and, in larger sizes, by use of variable-speed compressor motors, multiple cylinders with unloading or by-passing, or multiple compressors. Some of these methods may be relevant to the problem encountered with domestic units, in that motors above a certain size produce unacceptable starting transients on single phase supplies. By coupling two compressors together with common lubrication it may be possible to achieve double the output on single phase whilst still using relatively cheap components.[21]

4. APPLICATIONS OF HEAT PUMPS

As a result of rises in the cost of energy, capital-intensive energy conservation devices have become attractive options for space conditioning and process heating. However, the largest market for heat pumps at present (that in the USA) has developed because, in some cases, the gas utilities were unable to supply all their prospective customers. If coupled with restrictions on the use of direct electric heaters, this makes the case for using heat pumps very strong indeed.

4.1. Space and Water Heating in Buildings

The growth of the heat pump market in the USA has followed the increase in use of space cooling systems (air conditioning) in all classes of building. The market in colder climates, such as northern Europe, has been slower to develop since the heat pump must normally be sold purely on its heating ability.

As the same hardware can fulfil both heating and cooling functions, the heat pump offers a relatively economical method of heating, if the building also requires cooling. The heat pump for heating and cooling is slightly more complicated than the single function machine and extra safeguards must be incorporated.[22] The performance is not as good in either mode as a single function machine—as the prime requirement is for cooling this determines the size of the unit and it may therefore be undersized for heating. The fin spacing on such machines has already been mentioned and siting the compressor in the outdoor part is another cause of degradation of heating performance. The cooling function means that these machines are air-source units; air output and central heating are provided by warm air distribution (the most favoured system in new USA construction).

Initially, the approach in Europe was to follow American practice (air-to-air) but for heating-only machines. This is incompatible with the most common type of heating system (water radiator) and is difficult to apply retrospectively to existing construction. Air-to-water heat pumps for central heating and domestic hot water are now being produced plus some individual room heaters; ground-source and air-to-air units (using ventilation extract air as source) are more novel approaches. Domestic hot water is delivered at 55 °C which is low by conventional standards but is acceptable for most tasks, subject to sufficient storage volume being available.

The high levels of internal heat gain in commercial buildings provide opportunities for use of the cooling-with-heating heat pump. Although this may not produce the lowest running cost heating system the first cost advantages of a single rooftop package can be sufficient to warrant application (Fig. 8.5) but the benefits of the heat pump can be wasted by poor control—without suitable anticipation, excessive use may be made of supplementary heaters in preheating. Small cooling-with-heating heat pumps using internal (water) ring-mains provide a convenient method of

FIG. 8.5. A single package heat-pump unit mounted on the roof of a shop. (Reproduced by permission of the Electricity Council.)

transferring heat within a building—from an area where cooling is required to one which needs heating. Whilst this is useful for refurbishing buildings, greater energy efficiency may be achieved with central cooling plant supplying reclaimed heat via a separate distribution system. Only a small penalty will be paid for raising the output temperature sufficiently, but provision must be made for supplementing the heat input and rejecting excess heat (via a double-bundle condenser and a cooling tower) when the heating and cooling loads are not in balance.

4.2. Industrial Process Heating

As a result of recent development work[23] the output temperatures available for industrial use have been raised to 100 °C and it is hoped to go higher. This has been achieved by solving problems in compressor design, lubrication and oil stability and may require new refrigerants since the current ones, such as R-113, have critical temperatures around 200 °C. However, the available temperatures are sufficient for a number of industrial applications especially as the requirement for heat often implies a corresponding supply of waste heat at a usefully high temperature.[13]

The most closely matched source/sink pair is in the drying application, where air is recirculated over the product, then over the evaporator (to dehumidify) and the condenser (to reheat). This method is used for timber and ceramic drying and produces a more uniform product than conventional methods at higher PER even though electric motor-driven compressors are used.[24] An analogous method (mechanical vapour recompression) uses the process solvent as the working fluid to achieve concentration of the solution.

4.3. Heat Re-utilisation

The provision of mechanical ventilation in swimming pools to control moisture levels, especially condensation, is a virtually ideal application for heat pumps since the outgoing air is warm and humid and low temperature heat for the incoming air and pool and shower water is required. Relatively large plant is used for this task and values of COP_h in the region of 5–7 are obtained.[25]

Other applications are in similarly well ventilated buildings such as schools and hospitals, and reutilisation of heat from refrigeration equipment, such as bulk milk coolers.[26] The most effective method in some cases is the combination of heat exchanger and heat pump—the heat exchanger carrying the 'downhill' part of the heat recovery process and the heat pump making up the difference.

5. TRENDS AND FUTURE DEVELOPMENTS

Substantial technical advances are in prospect of which the most notable is the direct use of fossil fuels to drive heat pumps—although the overall efficiency of the heat pump cycle may not be any better than is now achieved, the possibility then exists for on-site recovery of waste heat and, by this means, the PER may be raised above that of any alternative process. Various lines are being followed including a gas-fired absorption cycle[27] and a Rankine power cycle coupled to a reversed-Rankine heat pump cycle[28] for domestic use. In larger buildings, internal combustion engine driven heat pumps would offer the same advantages,[29] and the maintenance requirements, which would be a disadvantage for the domestic user, could be adequately met. Quantity production of any such device should lead to improvements in cost and reliability over current one-off units. It is to be expected that the heat pump will become a much more common domestic product in future but the first wide scale acceptance of them will probably occur in commercial buildings.

ACKNOWLEDGEMENT

This chapter is based on work of the Building Research Establishment and permission to use it is acknowledged.

REFERENCES

1. ZEEMAN, M. W. *Heat and Thermodynamics* (5th edition), 1968, McGraw-Hill, New York.
2. DIDION, D. A. In: *Heat Pump and Space Conditioning Systems for the 1990s*, 1979, Carrier Corporation, New York, 37–47.
3. ROUVEL, L. *Heiz., Lüft., Haustech.*, 1975, **26,** 393–96.
4. ANON. *Heat., Vent. Eng.*, 1977, **51,** October, 24.
5. BIEHL, R. A. *ASHRAE J.*, 1977, **19**(7), 20–24.
6. KOLBUSZ, P. In: *Heat Pumps and their Contribution to Energy Conservation*, 1976, Noordhoff, Leyden.
7. KLEINPETER, M. *Elektrowärme Int.*, 1978, **36,** A125–27.
8. FREUND, P., LEACH, S. J. and SEYMOUR-WALKER, K. *Heat Pumps for Use in Buildings*, 1976, Building Research Establishment, Watford.
9. HEAP, R. D. *Elektrowärme Int.*, 1977, **35,** A77–81.
10. DIEDRICH, H. *Elektrowärme Int.*, 1976, **34,** A71–75.
11. RINCK, TH. In: *Wärmepumpen*, 1978, Vulkan–Verlag, Essen, 192–99.
12. BULLOCK, C. E. *ASHRAE J.*, 1978, **20**(9), 38–43.
13. MATTALON, R. W. In: *Proc. 8th Congress Union Internationale d'Electrothermie*, 1976, Liege.
14. KALISCHER, P. In: *Study Day on the Development of Heat Pumps in the Community for Heating and Air-conditioning*, 1978, Commission of the European Communities, Brussels.
15. VIELHABER, K. *Elektrowärme Int.*, 1977, **35,** A283–86.
16. KAYS, W. M. and LONDON, A. L. *Compact Heat Exchangers* 2nd Edn., 1964, McGraw-Hill, New York.
17. MATSUDA, T., MIYAMOTO, S. and MINOSHIMA, Y., *ASHRAE J.*, 1978, **20**(8), 32–35.
18. ANON. *Heat Pump TA3001*, 1976, Electricity Council Overseas Activities Translation 2044, London.
19. KELL, J. R. and MARTIN, P. L. *J. Inst. Heat Vent. Eng.*, 1963, **30,** 333–56.
20. STANNOW, J. C. In: *Wärmepumpen*, 1978, Vulkan–Verlag, Essen, 105–09.
21. STANNOW, J. C. In: *Heat Pump and Space Conditioning Systems for the 1990s*, 1979, Carrier Corporation, New York, 137–46.
22. ANON. *Report RP59*, 1971, Edison Electric Institute, New York.
23. HODGETT, D. L. *Improving the Efficiency of Drying Using Heat Pumps*, 1976, Electricity Council Research Centre, Capenhurst.
24. MORTGAT, J. H. *An Industrial Dryer Employing a Heat Pump for Plaster Blocks*, 1977, Electricite de France, Paris.
25. HAUSMAN, H. *Elektrowärme Int.*, 1979, **37,** A175–80.

26. Cromarty, A. S. *J. agric. Eng. Res.*, 1968, **13,** 225–40.
27. Sarkes, L. A., Nicholls, J. A. and Menzer, M. S. *ASHRAE J.*, 1977, **19**(3), 36–41.
28. Strong, D. T. G. In: *Antriebe für Wärmepumpen*, 1979, Vulkan–Verlag, Essen, 95–102.
29. Herbst, D. In: *Antriebe für Wärmepumpen*, 1979, Vulkan–Verlag, Essen, 134–40.

Chapter 9

HEAT EXCHANGERS FOR WASTE HEAT RECOVERY

D. A. Reay

International Research and Development Co. Ltd, Newcastle upon Tyne, UK

SUMMARY

Waste heat recovery is an area of application for heat recovery exchangers which is rapidly growing in importance as the cost of energy rises. This has led to major research, development and demonstration programmes in this field in many industrialised nations, not least in the UK.

The use of heat exchangers for heat recovery is normally associated with industrial processes and boiler plant. It does however extend to incinerators, prime movers, large air conditioning systems, and even into the home; air-to-air heat exchangers are being installed in many houses which employ forced ventilation heating systems.

While several heat exchangers used for heat recovery have been developed for general process heat exchange, some remain unique to heat recovery applications. These include regenerative and heat-pipe based systems, along with several other gas–gas types. A review of the many contenders in the field reveals several developments of interest which will ensure that this sector remains of prime importance.

1. INTRODUCTION

It may be argued that the design and use of heat exchangers for waste heat recovery employs technology common to many other areas of heat exchanger application, and to highlight this particular use may mislead the designer into over-estimating its importance. Indeed, one finds, particularly among the older-established equipment manufacturers (with the notable exception of those dealing with gas–gas heat exchangers), a feeling that heat recovery has been practised for many years using essentially standard equipment. There is also a belief, possibly justified in some instances, that companies who have entered the field more recently with product ranges directed specifically at the heating, ventilating, air conditioning (HVAC) and low-temperature (<300 °C) industrial process heat recovery fields, have sacrificed product quality, reliability and the ability to offer a full design, installation and commissioning service to the user.

The need to recover waste heat as a way of saving energy and minimising the operating costs of HVAC and industrial plant is, however, not in dispute, and the growing range of application areas involving conditions outside those normally encompassed by process heat exchange has led to a substantial growth in product development, directed both at increasing the efficiency of existing designs and designing new systems to meet fouling criteria, higher (or lower) process temperatures, etc.

It is, I believe, fair to say that the shell-and-tube heat exchanger has been less affected by this trend than any other type of system—general process requirements have led to performance improvements in their own right, as has been shown in previous chapters, and also, in the field of liquid–liquid heat recovery the same may be said of the plate heat exchanger.

Many heat exchangers have been developed solely for waste heat recovery applications, however; and it is these, particularly for use in exhaust gas streams, which will be concentrated on here. The equipment covered includes gas–gas, gas–liquid and liquid–liquid heat recovery units, with a bias towards systems for use in industrial processes, where the technological challenges are often as important as the economic challenges of HVAC applications (at least in the UK). Following discussion of the systems and important developments, the chapter concludes with a brief review of some of the significant present and future application areas. Occasional brief reference is also made to the heat pump, which is also of great significance in the context of waste heat recovery, but this is dealt with in detail in Chapter 8.

2. GAS–GAS HEAT RECOVERY EQUIPMENT

2.1. Introduction

Of the techniques for waste heat recovery, it is in the area of gas–gas heat exchange that the widest variety exists. However, commercially available systems may be broadly grouped into two classifications: recuperators and

2.8. Recuperative Burners

The use of convection or radiation recuperators for preheating combustion air has been common for many decades. However, it is only comparatively recently that the recuperator has been integrated into burner design, to produce recuperative burners, and this is of sufficient interest to warrant separate discussion.

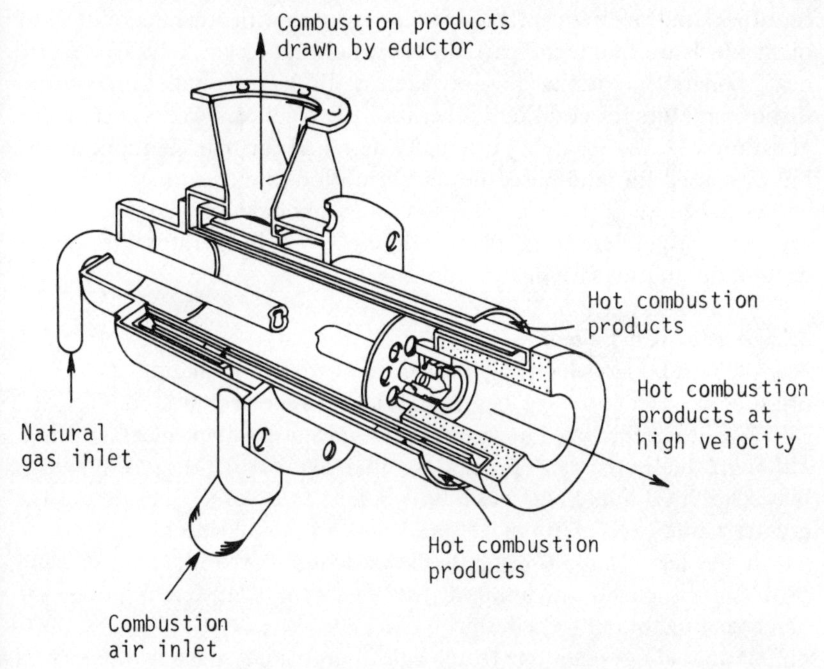

FIG. 9.8. A self-contained recuperative burner.

A recuperative burner developed by the British Gas Corporation[11] is shown in Fig. 9.8. It incorporates a counterflow heat exchanger in the form of concentric annuli around the burner nozzle, in which flue gases are used to preheat combustion air. Preheats of up to 650 °C are achievable, and this can lead to fuel savings of 30–40 %.

As with any exhaust-gas heat exchanger, materials selection is particularly important at high metal temperatures, and current developments by burner manufacturers are directed at achieving sustained operation with heavy fuel oil-fired units where, of course, corrosive products of combustion abound.

3. GAS–LIQUID HEAT RECOVERY EQUIPMENT

3.1. Introduction

The gas–gas heat recovery units described above may be considered in many applications as retrofitted items of equipment which, apart from saving energy, have little influence outside the immediate environment of the process to which the device is fitted. Gas–liquid heat recovery units, on the other hand, are frequently used in conjunction with other major items of plant which are an integral part of the overall heat recovery system. Waste heat boilers in particular are a case in point, linked to steam turbogenerators for electricity generation. Typical of such a system is the Hawthorn Leslie 'Seajoule', originally designed for marine applications, but now used for land-based power generation.

As will be seen later, the applications of gas–liquid heat recovery systems are very broad, covering plant ranging from incinerators to power generation and many industrial processes.

3.2. Waste Heat Boilers

A waste heat boiler, which may be natural or forced circulation, and of the fire-tube or water-tube type, bears little resemblance to its fired counterpart. (A waste heat boiler may, however, be provided with supplementary firing, although this is the exception rather than the rule.) Design differences between conventional and waste heat boilers arise largely because of the greater variety of environments, and hence potential problem areas, in which the latter have to operate. Dust loads, corrosion hazards, and problems associated with adapting the boilers to the upstream process are all critical factors in their design.

This should not deter one from neglecting to study the advantages of a waste heat boiler over, say, gas–gas heat recovery, however. Installations tend to be lower on capital cost than other heat recovery systems of comparable duty, partly because of the compactness of the waste heat boiler, aided by high heat transfer rates in the boiling process. This high rate of heat transfer also has advantages in that the metal temperatures are maintained fairly low (a greater advantage in very high-temperature applications where, of course, acid dew point problems may be avoided). Set against this, however, assuming that a use exists for high grade steam, is the need for some form of water treatment, as the feedwater for the boiler, whether it be directly fired or not, has to be of a certain quality.

With regard to sizes, waste heat boilers can be economically acceptable for duties as low as 500–100 kg steam per hour. Fire-tube boilers cover the range up to 20 000 kg steam per hour and operating pressures up to 20 bar. For higher duties and pressures, water-tube units would be selected, and the whole package may also incorporate economizers, superheaters and even an air preheater.[3]

Although it is currently expensive, the use of tubes having an internal porous boiling surface may have a role to play in waste heat boilers. Also, waste heat boilers are normally associated with the need to raise steam. A growing demand for systems capable of converting low grade heat, be it in a liquid or gaseous stream, into electricity will lead to 'boilers' for evaporating Freon for vapour turbines, also using high heat-flux tubing (see also Chapter 8).

3.3. Economizers

The economizer is commonly, and unfortunately, regarded as a waste heat recovery unit unique to boilers, although its application is spreading to other processes where hot exhaust gases may be used for heating water or other fluids.

Most economizers consist of a bundle of finned tubes spanning the exhaust gas duct (although on heavy fuel oil-fired units, the first few rows may be unfinned). A number of passes are commonly employed on the liquid side, and a single pass is used on the gas side. Economizers are normally constructed of steel, although originally cast iron was used, and the fins are generally welded to the tube. Construction is often much more substantial than on many other types of recuperative heat exchangers, partly because of the aggressive environments in which they operate in many boiler flues, and the susceptibility of tubes at the back end of the heat exchanger to suffer from corrosion. There are a number of ways in which corrosion, caused by condensation because of low water temperatures and hence low metal temperatures, can be prevented. Bunton[12] describes how a proportion of the economizer outlet water may be returned to the inlet to preheat the make-up, thus preventing condensation by virtue of the resulting higher back-end temperature. Tests are also being carried out on coated tubes, with a view to preventing sooting, hence improving heat transfer.

Economizer design and manufacturing practices have been developed by a comparatively small number of companies—e.g. Greens and Senior Economizers in the UK—and in my opinion, while we are unlikely to see dramatic changes in economizer design, the wealth of experience available from manufacturers and users of such plant is a much underrated asset. Fouling and corrosion bedevil all types of heat exchangers at one time or another—the many organisations launching new products in this field would do well to study the history of economizers.

3.4. Thermal Fluid Heaters

It may be argued that thermal fluid heaters[13] could, in the context of their use as waste heat recovery devices, be adequately categorised as economizers. This separate terminology, adapted primarily for the fired variety is, however, considered appropriate here as being the name under which such systems are marketed.

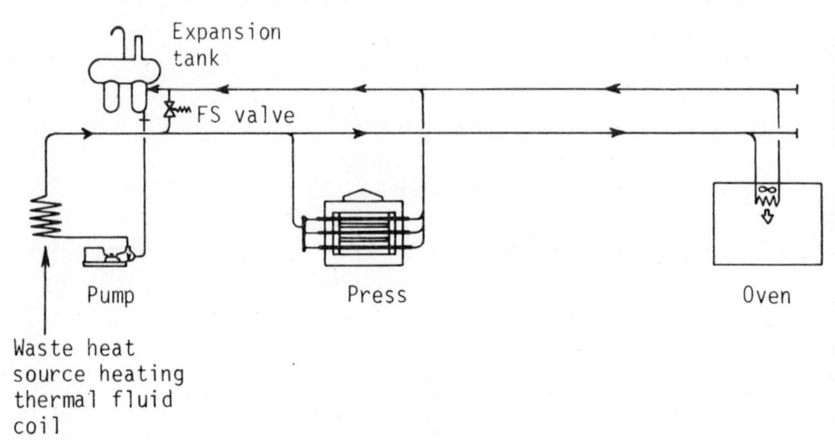

FIG. 9.9. Items of plant served by a thermal fluid loop, supplied by a waste heat source.

As well as using steam or high-pressure hot water as a transfer medium for heat, a thermal fluid, typically an organic such as a diphenyl–diphenyl oxide eutectic (Dowtherm A or Thermex), a mineral oil, or a salt, may be pumped around the circuit serving the various heat users, as shown in Fig. 9.9. The main advantage of the use of thermal fluids is the avoidance of the high pressures associated with steam or high-pressure hot water systems. Dowtherm A, for example, has an atmospheric boiling point of about 260 °C, and hence savings in installation cost can be significant. Boyen[13] cites the elimination of large-diameter vapour ducts, flash vessels, safety valves and pressure control devices as offering potential savings of between 25 and 50 % on installed cost. The reduction in maintenance due to the use of a non-pressurised system will lead to further savings.

The use of organic fluids in heat recovery installations is not restricted to single-phase heat transfer. The diphenyl–diphenyl oxide eutectic is also used in two-phase systems where heat uniformity is particularly important. In both phases, however, care should be taken to ensure that materials are compatible, and the maximum film temperatures recommended by the fluid manufacturer are not approached, thus ensuring that the working fluid has an acceptable service life.

These criteria to some extent determine the design of the gas–liquid heat exchanger used to heat the circulating fluid (normally exhaust gas rather than hot process liquid is the heat source used). As with a waste heat boiler, fire-tube or liquid-tube configurations are possible, the latter generally being adopted, as they offer better control of film temperatures and may be adapted directly as a waste heat unit from standard fired fluid heaters.

As with other heat exchanger complexes, the use of computer optimising routines is likely to lead to improved heat utilisation where a number of heat sources and sinks exist throughout a plant.

3.5. Fluidised-bed Units

There is no need here to describe in detail the process of fluidisation. In brief, a bed of particles with a gas flowing through it can, under certain

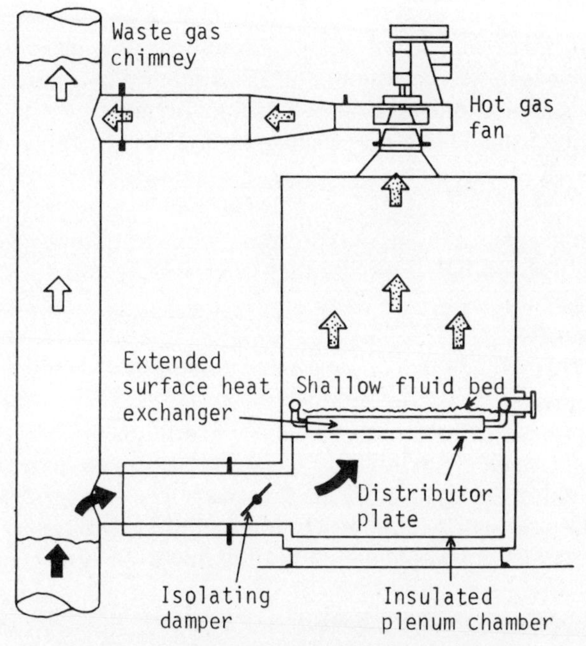

FIG. 9.10. A fluidised bed heat recovery unit.

conditions, take on fluid-like properties which are ideal for combustion and heat transfer.[14] It is in the area of combustion that most of the development effort on fluidised beds has been directed. However, the system is of substantial interest as a basis for waste heat recovery equipment, particularly gas-to-liquid.

Howard[14] cites experimental work at Aston University which demonstrated that 1000 m^2 of extended surface could be accommodated within 1 m^3 of shallow fluidised bed, and heat transfer coefficients of 4 kW/m^2 K could be obtained (based on the finned tube outside diameter). Use of a shallow bed ensures that the pressure drop is low.

Stone–Platt Fluid fire heat recovery units based on fluidised beds are commercially available, and a unit is shown schematically in Fig. 9.10. The fluid to be heated may be water or a thermal heat transfer fluid, and this particular system is available with outputs of between 75 kW and 1·5 MW. Other work at Stone–Platt has led to the falling-cloud heat exchanger, in which particles dropped down an exhaust gas flue, thus collecting heat, are subsequently passed to a fluidised bed where fluidisation encourages them to give up their heat to a heat exchanger immersed in the bed. The particles are then transported back to the top of the flue.

One of the main attractions of this, and other fluidised bed heat exchangers, is the ability of the unit to resist fouling, provided of course that the plate on which the bed is supported is maintained in a clean condition. The movement of a fluidised bed between a source and sink is a logical next step from the falling-cloud heat exchanger, and the general field is one of major growth potential.

4. LIQUID–LIQUID HEAT RECOVERY

4.1. Introduction

It is not proposed to dwell at length on the subject of liquid–liquid heat recovery, because one of the two main devices used in this area, the shell-and-tube heat exchanger, has been considered in detail elsewhere in this book. Also, the liquid–liquid heat exchanger, be it of this type or the plate form, is used routinely in most processes involving liquid flow, and its application to waste heat recovery follows similar guidelines to those laid down for solving any liquid–liquid heat exchange problem.

4.2. Shell-and-tube Heat Exchangers

The basic shell-and-tube heat exchanger has been developed over many years for process applications, and while improvements in performance are still being made, their effect is small.[15] Two areas where important developments have led to increased potential in heat recovery do exist, however. These are the development of units employing non-metallic materials, typically graphite and plastics, for use in acid plant and where other corrosive liquids are found, and replacement of conventional tube

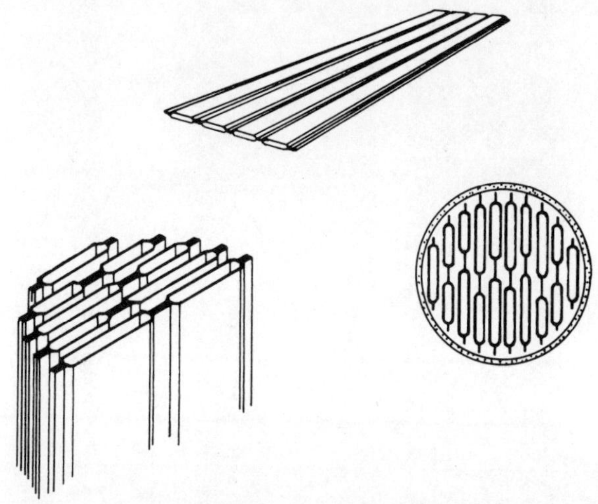

FIG. 9.11. 'Lamellae' in an Alfa-Laval liquid–liquid heat exchanger.

bundles within the shell by more advanced concepts, some of which are directed at achieving true countercurrent flow.

Representative of the latter is the lamellar heat exchanger marketed by Alfa-Laval. Illustrated in Fig. 9.11, the lamellae are flattened tubes made up using strips of plate seam-welded together. The space inside each lamella is ensured by forming of the strips, while external protruberances act as spaces on the shell side. This heat exchanger comes part of the way towards bridging the gap between the basic shell-and-tube unit and the plate heat exchanger, discussed below.

4.3. Plate Heat Exchangers

The plate heat exchanger, while subject to temperature and pressure constraints significantly lower than those for shell-and-tube heat exchangers, offers considerable advantages in terms of compactness and thermal efficiency, 90% heat recovery being regularly achieved with a unit of acceptable size and capital cost.

FIG. 9.12. APV Paraflow type R145. (By courtesy of The APV Company Ltd.)

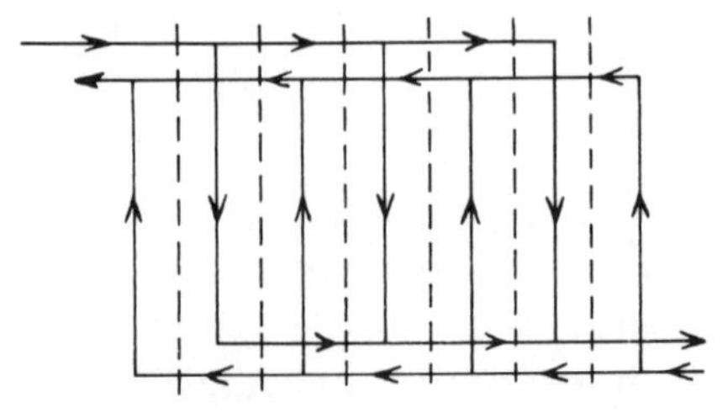

FIG. 9.13. Flow paths in a liquid–liquid plate heat exchanger.

Most plate heat exchangers consist of a frame in which metal plates are clamped together between a 'header' and 'follower', as shown in Fig. 9.12. The plates are corrugated to promote turbulence, and are sealed by gaskets along the edges and around corner ports. Plates are grouped into passes, with each fluid being directed evenly between the paralleled passages in each pass. A single pass counter-flow arrangement as illustrated in Fig. 9.13 is normally adopted.

Gasket materials limit maximum fluid temperatures to 200–250 °C at present, but plate materials may be selected from a wide choice, stainless

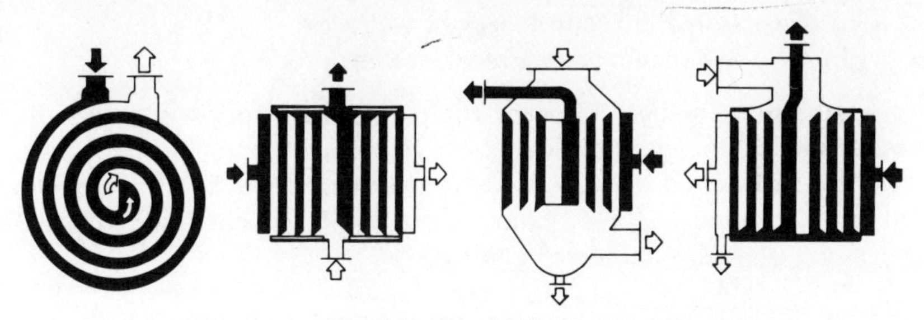

FIG. 9.14. The Alfa-Laval spiral heat exchanger.

steel being the most common. It has been argued[16] that the turbulence-inducing elements could be introduced into the gaskets, rather than the plates, saving on tooling and hence capital cost. This may however limit their acceptability in the food industry, where current construction lends itself to easy cleaning.

A second type of plate heat exchanger, used much less widely, is the spiral unit. Illustrated diagrammatically in Fig. 9.14, the spiral heat exchanger may be operated in true counter-flow, and while limited, like the conventional plate heat exchanger, to fairly low maximum pressures, typically 10–15 bar, the welded structure permits somewhat higher operating temperatures.

5. APPLICATION AREAS

The application of waste heat recovery equipment covers all fields of heat utilisation, and, because of the rising cost and potential shortages of fuel, is growing considerably within each of these fields. Four main application areas will be discussed here: (1) heating, ventilating and air conditioning, (2) industrial processes, (3) incinerators and (4) prime movers.

Within these broad application areas, some, or even all, of the ways in which heat can be 'lost' (and therefore potentially 'recovered') may exist. These include:[17]

(i) Heat in flue gases.
(ii) Heat in excess air associated with the combustion process.
(iii) Heat in excess air associated with the process or equipment.
(iv) Heat loss from the exterior of the equipment.
(v) Heat losses in cooling water.
(vi) Heat losses in providing chilled cooling water.
(vii) Heat stored in products leaving a process.
(viii) Heat in vapours and gases leaving the process.

Applications of heat recovery equipment are strongly governed by economic factors, and this extends through domestic to industrial fields. The 'payback period' must be attractive—currently in the UK, for example, industrial heat recovery installations having payback periods in excess of 3 years are generally considered unattractive.

5.1. Heat Recovery in Heating, Ventilating and Air Conditioning

The field of HVAC has long been a user of air–air heat recovery equipment, in particular the rotating regenerator and the run-around coil. More recently both plate and heat-pipe heat exchangers have been applied and, of course, the heat pump has had a major role to play in, for example, simultaneous heating and cooling of different areas of a building.[18]

Hygroscopic heat wheels are particularly attractive in HVAC applications. While, in common with other heat exchangers, they are able to precool incoming air in summer as well as providing preheat in winter, they are also able to transfer moisture. This humidity exchange is required either to dehumidify the supply air in summer or to increase the humidity in winter. Where year-round operation can be justified, savings in investment in humidifying plant and, more particularly, in the chiller capacity, can be even more dramatic than the value of energy savings due to heat recovery. (This is of course primarily applicable only when the unit is designed into the overall system for the building during planning stages.) It is interesting to note that in some countries, Sweden for example, the use of heat recovery equipment on air-handling units in larger buildings is now mandatory.

Waste heat recovery in the home may be regarded by some with scepticism. However, under certain circumstances a serious case can be made for air–air heat recovery units, several of which are manufactured in Europe and routinely used in Scandinavia. In the UK, for example, condensation is a serious problem in many houses, created partly by lack of ventilation, and partly by occupiers conserving energy in order to keep fuel costs to a minimum. One technique being tested for overcoming condensation is to incorporate forced ventilation systems with a primary exhaust duct into which heat recovery equipment may be installed. The recovered heat is used to preheat fresh inlet air.

5.2. Process Heat Recovery in Industry

Industrial heat recovery has been chosen by the Department of Energy in the UK for priority action, in part because the potential yearly savings are substantial—of the order of 5–10 mtce (million tons of coal equivalent)[19]—equivalent to the whole annual process energy use of the brick, glass and cement industries. The applications embrace all the types of equipment described above, and more, and there is considerable incentive to improve heat exchanger performance, particularly with regard to fouling, dew point and installed cost.

It is generally preferable, more especially in the case of gas–gas systems, to use recovered heat in the process where it originates. As well as ensuring a minimum installation cost, due to the often close proximity of heat source and sink, the heat demand will always coincide with waste heat availability. Also, of considerable economic importance, the recovered heat, if used in a process, can give benefits throughout the year, whereas if used for space heating the demand is seasonal. In applications involving heat recovery from liquid effluents, use of the heat in another process or, for example, preheating boiler feedwater is very common. (It is worth noting that heat pumps have a role to play here.) Payback periods on installations are typically 1–3 years, although they tend to be somewhat longer on heat pumps.

Fouling is a major problem on many potential heat recovery installations, and one way in which this is being tackled is by using disposable heat exchanger elements. Introduced in Switzerland, and based on the plate gas-to-gas heat exchanger, such units may be used economically in dryers, annealing kilns and other areas where contaminated exhausts exist. Payback periods of 12 months, using new heat exchanger elements every 3 months, have been demonstrated.

5.3. Other Application Areas

Heat recovery is important in two other major fields. In incinerators, the use of waste heat boilers is becoming popular for steam raising for electricity generation, or for district heating. The growth of legislation in the field of pollution control has also spurred developments in the field of high-temperature heat recovery. At present of greater importance in North America, legislation on fume incineration requires that industrial process fumes be incinerated at, say, 700–800 °C. This may necessitate after-burning, which can be expensive in terms of fuel, and economic sense forces the adoption of rotating regenerators, tubular recuperators, or waste heat boilers downstream of these incinerators. In cases where the pollutant itself can be combusted using a catalyst, heat exchangers combining the role of incinerator and heat recovery unit are under development.

Prime movers, either as sources of waste heat, or, for example, in the case of organic cycle turbines users of waste heat, are also of interest. Growing interest in diesel or gas engine-driven heat pumps, which are only justifiable economically if the heat from the engine is recovered, has given impetus to exhaust waste heat boiler development—much of course can be learnt from the marine field.

In the context of marine engines, it is of significance to note that as the thermal efficiency increases, exhaust gas temperatures decrease. A time is foreseen when exhaust heat recovery will only be justified if the exhaust is brought below the acid dew point by the heat recovery unit—this poses a materials design challenge which could benefit many other fields of heat exchanger application.

In my opinion, the next decade will see a switch away from conventional heat exchangers towards heat pumps, in any heat recovery application where dehumidification of the exhaust could permit recycling, i.e. dryers. Process improvements should also make the need for heat recovery less great—although that is even further ahead. Meanwhile, the industrial process application area in particular remains one of steady growth, with many interesting developments underway.

REFERENCES

1. Sohlberg, J. *Heizung Luft Klimatech.*, 1976, **27**(3), 18.
2. Chojnowski, B. and Chew, P. E. *CEGB Research*, 1978, **7**, 14–21.
3. Reay, D. A. *Heat Recovery Systems. A Directory of Equipment and Techniques*, 1979, E. and F. N. Spon, London.
4. Strindehag, O. *Build. Serv. Eng.*, 1975, **43**, 52–56.
5. Dugwell, D. R. *Steel Times Ann. Rev.*, 1977, **7**, 784–93.
6. Dunn, P. D. and Reay, D. A. *Heat Pipes*, 2nd Edn., 1978, Pergamon Press, Oxford.
7. Reay, D. A. *Heat. Vent. Eng.*, 1979, **53**, 8–14.

8. Ranken, W. A. and Lundberg, L. B. Proc. 3rd Int. Heat Pipe Conf., AIAA Paper 78-435, 1978, AIAA, Palo Alto.
9. Ruch, M. A. and Grover, G. M. Proc. 2nd Int. Heat Pipe Conf., ESA Report SP112, 1976, (1).
10. Moritz, K. *Chemie Ing. Technik*, 1969, **41,** 37–40.
11. Proffitt, R. *Gas Eng. Management*, 1977, **17,** 180–94.
12. Bunton, J. F. Proc. Inst. Plant Eng., Waste Heat Recovery Conference, 25–26 Sept., 1974, London.
13. Boyen, J. L. *Practical Heat Recovery*, 1975, John Wiley and Sons, New York.
14. Howard, J. R. In: *Energy for Industry*, (ed. P. W. O'Callaghan), 1979, Pergamon Press, Oxford, 111–29.
15. Bergles, A. E., Webb, R. L. and Junkan, G. H. *Energy*, 1979, **4,** 193–200.
16. Klaschka, J. T. In: *Energy for Industry*, (ed. P. W. O'Callaghan), 1979, Pergamon Press, Oxford, 277–95.
17. Anon. The recovery of waste heat from industrial processes, Fuel Efficiency Booklet 13, 1978, Department of Energy, London.
18. Reay, D. A. and Macmichael, D. B. A. *Heat Pumps. Theory, Design and Applications*, 1979, Pergamon Press, Oxford.
19. Anon. Energy conservation research, development and demonstration, Energy Paper Number 32, 1978, HMSO, London.

BIBLIOGRAPHY

1. Abuaf, N., Jones, O. C., Jr. and Zimmer, G. A. Optical probe for local void fraction and interface velocity measurements, BNL-NUREG-50791, 1978, Brookhaven National Laboratory, Upton, New York.
2. Abuaf, N., Jones, O. C., Jr., Zimmer, G. A., *et al.* BNL flashing experiments: test facility and measurement techniques, BNL-NUREG-24336, 1978, Brookhaven National Laboratory, Upton, New York.
3. Ageev, A. I., Belushkin, V. A., Zel'dovich, A. G., *et al.* Analysis of cooling and cryostating processes for large superconducting facilities. (In Russian.), JINR-R-8-10039, 1976, Jt. Inst. Nucl. Res., Lab, High Energy, Dubna, USSR.
4. Aihara, T., Taga, M. and Haraguchi, T. Heat transfer from a uniform heat flux wedge in air–water mist flows, *Int. J. Heat Mass Trans.*, 1979, **22**(1), 51–60.
5. Alfa-Laval. *Heat Exchanger Guide*. The Alfa-Laval/De Laval Group, Order No. 60122, Reg. No. 350, Date 7103.
6. Allueva, C. and Demestre, J. M. Stainless steels and alloys in the nuclear industry. Corrosion resistance and mechanical properties. (In Spanish.) *Corros. Prot.*, 1978, **9**(5–6), 29–35.
7. Alvi, S. H., Sridharan, K. and Rao, N. S. L. Loss characteristics of orifices and nozzles, *J. Fluids Eng.*, Sept. 1978, **100,** 299–307.
8. American Society of Heating, Refrigerating and Air Conditioning Engineers. Methods of testing for rating unitary air conditioning and heat pump equipment, 1978, ASHRAE Standard 37-1978, ASHRAE, New York.
9. American Society of Heating, Refrigerating and Air Conditioning Engineers. Methods of testing air-to-air heat exchangers, 1978, ASHRAE Standard 84-1978, ASHRAE, New York.
10. Anderson, J. H. Plate-fin heat exchanger, US Patent 4,139,054. Appl. 28 Oct., 1977. Published 13 Feb., 1979.
11. Andersson, A. C. Thermal bridges in external walls provided with additional insulation. (In Swedish.) R46, 1978, Building Research Council, Sweden.
12. Andreev, V. K., Deev, V. I., Petrovichev, V. I., *et al.* Heat transfer conditions during bubble boiling of helium in a large volume. (In Russian.) *Teplofiz. Vys. Temp.*, 1978, **16**(4), 882–84.

13. ANDREEVA, A. B., ZABELIN, A. I., KOBZAR, I. G., *et al.* Corrosion of structural materials in the circuit of a nuclear power plant with a VK-50 reactor and effect of this process on the radioactivity of the coolant and deposits. (In Russian.) *Sb. Dokl. Seminara. Perspektivy Ispol'z, Yader. Reaktorov dlya. Teplosnabzh. Gorodov i Prom.* Predpriyatii Dimitrovgrad, 1977 (Pub. 1978), 108–20.
14. ANON. Heat transfer characteristics of horizontal tubes. Multiple effect evaporators. (In Japanese.) *Do*, 1978, **19,** 7–11.
15. AUNAAS, P. Significant savings using heat pumps in the new Asko building at Lillesand. (In Norwegian.) *Norsk VVS*, 1978, **21**(10), 825–27.
16. AUSSENAC, D., DOMENECH, S. and ENJALBERT, M. Production of fresh water by desalination of sea water. (In Spanish.) *Ing. Quim.* (*Madrid*), 1977, **9**(101), 97–103.
17. AVERIN, I. B. Heat transfer during turbulent flow of dissociating nitrogen dioxide in a channel. (In Russian.) *Izv. Vyssh. Uchebn. Zaved., Mashinostr.*, 1978, (9), 72–76.

18. BACHMANN, D. The Templifier—a heat pump for industry. (In German.) *Klima Kalte Ing.*, 1978, **6**(7–8), 257–59.
19. BALAKLEEVSKII, YU. I. and CHEKHOVICH, V. YU. Condensation of an immersed vapor jet. (In Russian.) *Teploperedacha pri Kipenii i Kondensatsii*, 1978, 77–93.
20. BANERJEE, S., YUEN, P. and VANDENBROEK, M. A. Calibration of a fast neutron scattering technique for measurement of void fraction in rod bundles, *J. Heat Trans.*, May 1979, **101**(2), 295–99.
21. BANKOFF, S. G. (ed.) *Topics in two-phase heat transfer and flow*, 1978, A.S.M.E., New York.
22. BASCO. Equa-Sperse fin tube tank heaters. American Precision Industries Inc., Bulletin 450, 2777 Walden Avenue, Buffalo, New York 14225.
23. BASU, D. X. and PINDER, K. L. Instantaneous heat transfer area in an evaporating drop, *Proc. 4th Nat. Heat Mass Transfer Conf.*, 1977, Sarita Prakashan, Meerut, India.
24. BAUER, B. A possibility for the more efficient use of heat to drive refrigeration plant and heat pumps. (In German.) *Brennst.-Warme-Kraft*, 1978, **30**(7), 292–96.
25. BAUMEISTER, K. J. and PAPELL, S. S. Geometrical correction factors for heat flux meters, NASA-TM-X-71560, 1974, N74-27878, Lewis Research Center, Cleveland, Ohio.
26. BAXI, C. B., BURHOP, C. J. and BENNETT, F. G. COBRA*GCFR, a computer code for thermal-hydraulic analysis of GCFR fuel assembly, *Trans. Am. Nucl. Soc.*, 1978, **30,** 543-45.
27. BEGOVICH, J. M. Hydrodynamics of three-phase fluidized beds, ORNL/TM-6448, Thesis, July 1978, University of Tennessee, Knoxville, Tennessee.
28. BELAN, N. V., BEZRUCHKO, K. V. and MATUSEVICH, V. A. Some problems of identifying the mathematical model of a two-phase spiral heat exchanger with an organic liquid as the working fluid. (In Russian.) In: *Dissotsiiruyushchie Gazy Teplonositeli Rab. Tela Energ. Ustanovok*, Volume 2, 1976, A. K.

Krasin (ed.), Akad. Nauk BSSR, Inst. Teplo-Massoobmena, MINSK, USSR, 129–35.

29. Belyakov, V. P., Shaposhnikov, V. A., Gorbachev, S. P., *et al.* Studies on nucleate boiling crisis of helium-I in channels of superconducting magnet systems. *IEEE Trans. Magn.*, 1979, MAG-**15**(1), 40–45.
30. Benedict, R. P. and Wyler, J. S. Analytical and experimental studies of ASME flow nozzles. *J. Fluids Eng.*, Sept. 1978, **100**, 265–74, (discussion 274–75).
31. Benvenuto, G. and Scotti, A. On the determination of the dryness fraction in wet steam flows by means of a calorimetric probe. (In Italian.) *Termotecnica*, 1978, **32**(10), 535–43.
32. Berbenni, P. and Spigai, G. Water in industrial cooling circuits. (In Italian.) *Inquinamento*, 1978, **20**(9), 43–49.
33. Beyer, G. Determination of geometrical and thermal criteria for cylindrical stirred vessels, heat exchangers, and reactors. (In Czech.) *Chem. Prum.*, 1978, **28**(8), 387–93.
34. Bezrodnyi, M. K. Disturbance of the stability of heat and mass transfer processes in some gas–liquid systems. (In Russian.) *Inzh.-Fiz. Zh.*, 1978, **34**(6), 1001–6.
35. Bezrodnyi, M. K. Generalization of experimental data on critical heat flows in the large-volume boiling of liquids. (In Russian.) *Izv. Vyssh. Uchebn. Zaved., Energ.*, 1978, **21**(11), 83–87.
36. Bienert, W. and Wolf, D. Liquid metal heat pipes for the central solar receiver, Intersociety 13th Energy Conversion Engineering Conference, San Diego, Calif., Proceedings, Volume 2, 20–25 August, 1978, Society of Automotive Engineers, Inc., Warrendale, Pa., 1040–49.
37. Bird, S. P. Uncertainty analysis routine for the ocean thermal energy conversion (OTEC) biofouling measurement device and data reduction procedure, PNL-2631, March 1978, Battelle Pacific Northwest Labs., Richland, Wash.
38. Birniehill Institute. *An introduction to industrial heat exchangers*, Birniehill Institute, 18–19 Sept., 1973. Course paper No. 2. East Kilbride, Glasgow: National Engineering Laboratory, Birniehill Institute, 1973.
39. Birniehill Institute. *Analysis, design and manufacture of heat pipes*, Birniehill Institute, 19–21 June, 1973. Course paper No. 1. East Kilbride, Glasgow: National Engineering Laboratory, Birniehill Institute, 1973.
40. Birniehill Institute. *Elements of two-phase flow*, Birniehill Institute, 13–14 Nov., 1973. Course lecture No. 1. East Kilbride, Glasgow: National Engineering Laboratory, Birniehill Institute, 1973.
41. Birniehill Institute. *Condensation and condensers*, Birniehill Institute, 11–15 Sept., 1972. Vol. 1, Course lecture notes No. 5. East Kilbride, Glasgow: National Engineering Laboratory, Birniehill Institute, 1972.
42. Biryukova, L. V., Danilov, N. N. and Sinitsyn, E. N. Superheating of pure liquids and binary solutions during convective heat exchange with a thin wire. (In Russian.) In: *Teplofiz. Metastab. Sist.*, 1977, P. A. Pavlov (ed.), Akad. Nauk SSSR, Ural. Nauchn. Tsentr., Sverdlovsk, USSR, 16–22.
43. Blander, M. Bubble nucleation in liquids, *Adv. Colloid Interface Sci.*, 1979, **10**, 1–32.

44. BOBROVA, G. I. and MORGUN, V. A. Experimental study of heat transfer in a porous heat exchanger. (In Russian.) In: *Nizkotemp. Tepl. Truby Poristye Teploobmenniki*, L. L. Vasil'ev (ed.), 1977. Akad. Nauk BSSR, Inst. Teplo-Massoobmena, im. A. V. Lykova, Minsk, USSR, 127–31.
45. BOLOTNIKOV, E. S. and BURUKHINA, L. E. Study of air drying using a lithium chloride solution. (In Russian.) *Zh. Prikl. Khim.*, 1978, **51**(9), 2084–86.
46. BONACINA, C., COMINI, G. and DEL GIUDICE, S. Evaporation of atomized liquids on hot surfaces. (In Italian.) *Termotecnica*, 1978, **32**(9), 472–81.
47. BOOM, R. W., EL-WAKIL, M., MCINTOSH, G. E., *et al.* Experimental investigation of the helium I two-phase flow pressure drop characteristics in vertical tubes, *Proc. 7th Int. Cryog. Eng. Conf.*, 1978, 468–73.
48. BORISHANSKII, V. M., ZHOKHOV, K. A., SVETLOVA, L. S., *et al.* Heat transfer in the movement of potassium in a tube in the supercritical region. (In Russian.) *Teplofiz. Vys. Temp.*, 1978, **16**(6), 1264–68.
49. BORODULYA, V. A., VINOGRADOV, L. M., GANZHA, V. L., *et al.* (eds.), *Heat and Mass Transfer in Multiphase Multi-component Systems.* (In Russian.) 1978, Akad. Nauk Beloruss, SSR, Inst. Teplo Massoobmena, Minsk, USSR.
50. BRADNER, J. The gas engine applied to heat pump installations. (In German.) *Kalte*, 1978, **31**(10), 458–66 and 471–72.
51. BRAND, B. A. Water treatment for district heating. Part 1, *Heat Air Condit. J.*, March 1979, **49**(566), 12, 14, 16 and 18.
52. BRAHAM, G. D. Heat pumps for industrial and commercial use, *Elect. Rev.*, 8 Dec., 1978, **203**(22), 37–39.
53. BREEV, V. V., GUBAREV, A. V. and PANCHENKO, V. P. Calculation study and projected development of energy installations with magnetohydrodynamic generators. (In Russian.) *Itogi Nauki Tekh.: Gener. Pryamogo Preobraz. Tepl. Khim. Energ. Elektr.*, 1978, **4**, 79–108, 128–43.
54. BRITISH STANDARDS INSTITUTION. Glossary of refrigeration, heating, ventilating and air conditioning terms, BS 5643, 1979, British Standards Institution, London.
55. BROKAW, R. S. Calculation of flue losses for high-efficiency furnaces and appliances, *ASHRAE J.*, 1979, **21**(1), 49 and 51.
56. BROUSSE, E. Study of a system using a heat pump and a 72 kW distillation column. (In French.) *Rev. Gen. Therm.*, Nov. 1978, **17**(203), 847–66.
57. BUGL, J. Nuclear power plants—design and operation. (In German.) *Schweissen Schneiden*, 1978, **30**(11), 446–50.
58. BUGLAEV, V. T. and TATARINTSEVA, T. I. Study of the effect of salt deposits in apparatus on heat transfer. (In Russian.) *Izv. Vyssh. Uchebn. Zaved., Energ.*, 1978, **21**(10), 81–85.
59. BUKIN, V. G., DANILOVA, G. N. and DJUNDIN, V. A. Heat exchange during evaporation and boiling of mixtures of refrigerants and oil in film evaporators. (In Russian.) *Holod. Tehn.*, 1977, (1), 33–36.
60. BULL, J. L. Graphite heat exchanger plant, *Process Eng.*, June 1969.
61. BULLOCK, C. Thermostat setback and residential heat pumps, *ASHRAE J.*, 1978, **20**(9), 38–43.
62. BUNDITKUL, S. and YANG, W-J. Laminar transport phenomena in parallel channels with a short flow construction, *J. Heat Trans.*, May 1979, **101**(2), 217–21.

63. BURDUKOV, A. P., KUVSHINOV, G. G. and NAKORYAKOV, V. E. Characteristics of heat transfer during nucleate boiling of an underheated liquid in a large volume at subatmospheric pressures. (In Russian.) *Teploperedacha pri Kipenii i Kondensatsii*, 1978, 94–118.
64. BURNSIDE, B. M. Gas turbine exhaust heat recovery by using the immiscible liquid binary cycle, *Termotecnica*, 1979, **33**(1), 11–19.
65. BUSSE, F. H. Non-linear properties of thermal convection, *Rep. Progr. Phys.*, Dec. 1978, **41**, 1929–67.
66. BYKOV, A. V., KROTKOV, V. N. and SUTYRINA, T. M. Intensification of freon air-cooled condensers due to new fin design, *Sci. Tech. Froid*, 1977, (4), 115–20.

67. CAMPANILE, A. Diesel-driven heat pumps. (In German.) *Klima Kalte Ing.*, 1978, **6**(11), 407–8.
68. CAPONE, G. J. and PARK, E. L., JR. Film boiling of Freon 113, normal pentane, cyclopentane and benzene from cylindrical surfaces at moderate pressures. *Int. J. Heat Mass Trans.*, 1979, **22**(1), 121–29.
69. CARTER, W. A., BUENING, H. J. and HUNTER, S. C. Emission reduction on two industrial boilers with major combustion modifications, KVB, Inc., 1978. PB 283 109, EPA/600/7-78/099A, KVB-6004-734. Tustin, Calif.
70. CHEMICAL ENGINEERING PROGRESS. Optimum design of multi-stage drying systems, *Chem. Eng. Prog.*, 1979, **75**(4), 61–65.
71. CHEPURNENKO, V. P., KIRILLOV, V. KH. and DYMOV, M. I. Design calculation of devices for treating air before air condensers of refrigerating units. (In Russian.) *Kholod. Tekh. Tekhnol.*, 1978, **26**, 89–94.
72. CHEREMISINOFF, N. P. and DAVIS, E. J. Stratified turbulent–turbulent gas–liquid flow, *AIChE J.*, 1979, **25**(1), 48–56.
73. CHIOU, J. P. Thermal performance deterioration of a crossflow heat exchanger due to flow non-uniformity, *Trans. ASME*, Nov. 1978, **100**(4), 580–87.
74. CHISHOLM, D. Prediction of pressure drop at pipe fittings during two-phase flow, 13th Internat. Institute of Refrigeration Congress, Volume 2, 27 Aug.–3 Sept., 1971, Washington, 781–89.
75. CHISHOLM, D. Air cooler, cooling towers and evaporative coolers. In: *An Introduction to Industrial Heat Exchangers*. Birniehill Institute, 18–19 Sept., 1973. Course Paper No. 4. East Kilbride, Glasgow: National Engineering Laboratory, Birniehill Institute, 1973.
76. CHISHOLM, D. Void fracture during two-phase flow, *J. mech. Eng. Sci.*, June 1973, **15**(3), 235–36.
77. CHISHOLM, D. Influence of pipe surface roughness on friction pressure gradient during two-phase flow, *J. mech. Eng. Sci.*, 1978, **20**(6), 353–54.
78. CHISHOLM, D. Heat pipes with header and artery systems, *Int. J. Heat Mass Trans.*, 1978, **21**(9), 1207–12.
79. CHISHOLM, D., DRUMMOND, W. and MURRAY, I. *Industrial Heat Exchangers*. Lecture Series, 3.1–3.9. 24–28 January, 1972. Von Karman Institute, Brussels.
80. CHUMAK, I. G., KOKHANSKIJ, A. I. and KUZNETSOVA, L. P. New method for calculating evaporative condensers. (In Russian.) *Izvest. vys. ucebn. Zaved. Energet.*, 1977, **20**(6), 79–85.

81. Chung, J. N. and Ayyaswamy, P. S. Material transport with steam condensation on a moving spray droplet including the effect of internal chemical reaction. In: *Top. Two-Phase Heat Transfer Flow, 1978 ASME Winter Annual Meeting*, S. G. Bankoff (ed.), 1978, ASME, New York, N.Y., 153–64.
82. Chung, J. N. and Ayyaswamy, P. S. Laminar condensation heat and mass transfer in the vicinity of the forward stagnation point of a spherical droplet translating in a ternary mixture: numerical and asymptotic solutions, *Int. J. Heat Mass Trans.*, 1978, **21**(10), 1309–24.
83. Cistjakov, F. M., Frolova, N. I. and Kuvsinov, S. G. Determination of the pressure drop in horizontal shell-and-tube evaporators with refrigerant boiling inside the tubes. (In Russian.) *Holod. Tehn.*, 1977, (2), 20–24.
84. Clyde, R. A. Extended area ceramic heat exchanger. In: *Alternative Energy Sources*, (*Proc. Miami Int. Conf.* 1977), T. N. Veziroglu (ed.), 1978, Volume 9, Hemisphere, Washington, D.C., 3981–87.
85. Coli, G. Protecting boilers against corrosion caused by fumes. (In Italian.) *Condiz. dell'Aria.*, 1978, **22**(8), 573–82.
86. Collins, R. L. and Lovelace, R. B. Experimental study of two-phase propane expanded through the Ranque–Hilsch tube, *J. Heat Trans.*, May 1979, **101**(2), 300–5.
87. Combs, S. K. Experimental study of ammonia condensation on vertical fluted tubes, CONF-780408-2, 6 April 1978, Southeastern Seminar on Thermal Sciences, Raleigh, N.C.
88. Combustion Engineering, Inc. Industrial application of fluidized-bed combustion, superheated steam boiler, Quarterly Report, 1 Oct.–31 Dec. 1977, HCP/T2473-18, March 1978, Combustion Engineering, Inc., Windsor, Conn.
89. Comfort, W. J. III, Alger, T. W., Giedt, W. H., *et al.* Calculation of two-phase dispersed droplet-in-vapor flows including normal shock waves, *J. Fluids Eng.*, Sept. 1978, **100,** 355–62.
90. Cooper, K. W. Saving energy with refrigeration, *ASHRAE J.*, 1978, **20**(12), 23–27.
91. Copin, S. R., Elufimova, S. M., Kozhevnikova, V. P., *et al.* Arrangement of a regenerative heat exchanger and the thermometer bulb of a thermo-regulating valve in commercial refrigeration equipment. (In Russian.) *Kholod. Tekh.*, 1978, (9), 31–34.
92. Costello, V. A., Melsheimer, S. S. and Edie, D. D. Heat transfer and calorimetric studies of a direct-contact latent heat energy storage system, In: *Therm. Storage Heat Transfer Sol. Energy Syst., 1978 ASME Winter Annual Meeting*, F. Krieth, R. Boehm and J. Mitchell (eds.), 1978, ASME, New York, N.Y., 51–60.
93. Craig, A. F. Double-pipe chillers and exchangers, Brochure AG6/3069, 1969, A. F. Craig Ltd, Paisley, UK.
94. Croix, J. M. Dropwise condensation with organic promoters. Effect of dripping onto a pipe bundle. (In French.) *Inf. Chim.*, 1978, **184,** 117–20.
95. Cur, N. and Sparrow, E. M. Measurements of developing and fully developed heat transfer coefficients along a periodically interrupted surface, *J. Heat Trans.*, May 1979, **101**(2), 211–16.

96. Dakin, J. T. Vaporization of water films in rotating radial pipes, *Int. J. Heat Mass Trans.*, 1978, **21**(10), 1325–32.
97. Davison, R. M. and Miska, K. H. Stainless-steel heat exchangers. Part 1. *Chem. Eng.* (*N.Y.*), 1979, **86**(4), 129–33.
98. Davison, R. M. and Mishka, K. H. Stainless-steel heat exchangers. Part 2. *Chem. Eng.* (*N.Y.*), 1979, **86**(6), 111–14.
99. De Maerteleire, E. Calculation of the external heat transfer coefficient in a helical cooling coil in an agitated Newtonian fluid. (In Dutch.) *Meded. Fac. Landbouwwet, Rijksuniv. Gent*, 1977, **42**(3–4), 2005–20.
100. Del Giudice, S., Comini, G. and Mikhailov, M. D. Finite element analysis of combined free and forced convection, *Int. J. Heat Mass Trans.*, Dec. 1978, **21**, 1619–21.
101. Department of Industry, Committee on Corrosion. Industrial corrosion monitoring, 1978, H.M.S.O., London, ISBN 0-11-512188-9.
102. Diamant, R. M. E. Condensation in buildings. Part 1. Basic principles, *Heat Air Condit. J.*, Feb. 1979, **49**(565), 42, 44, 46 and 48.
103. Diamant, R. M. E. Condensation in buildings. Part 2. Glazing, *Heat Air Condit. J.*, March 1979, **49**(566), 44, 46, 48 and 50.
104. Diamant, R. M. E. Condensation in buildings. Part 3. Moisture diffusion, *Heat Air Condit. J.*, April 1979, **49**(567), 24, 26 and 28.
105. Diamant, R. M. E. Condensation in buildings. Part 4. Condensation in roofs, *Heat Air Condit. J.*, May 1979, **49**(568), 30, 32 and 34.
106. Divilio, R. J. and Reed, R. R. Fly ash combustion in a fluidized-bed boiler, *AIChE Symp. Ser.*, 1978, **176,** 213–17.
107. D'Jackov, F. N. Heat exchange and hydraulic characteristics for boiling R-22 in internally finned tubes. (In Russian.) *Holod. Tehn.*, 1977, (1), 36–42.
108. Doichev, K. Some aspects of bubble formation in fluidized systems. (In Bulgarian.) *God. Vissh. Khim.-Tekhnol. Inst.*, *Sofia*, 1978, **23**(3), 133–40.
109. Dolgirev, Yu. E., Gerasimov, Yu. F., Maidanik, Yu. F., *et al.* Design calculation of a heat pipe with separate channels for steam and liquid. (In Russian.) *Inzh.-Fiz. Zh.*, 1978, **34**(6), 988–93.
110. Donadono, S. and Massimilla, L. Mechanisms of momentum and heat transfer between gas jets and fluidized beds, In: *Fluid.*, *Proc. 2nd Eng. Found. Conf.*, J. F. Davidson and D. L. Keairns (eds.), 1978, Cambridge Univ. Press, London, 375–80.
111. Dornier-system. Design and development of a gas controlled heat pipe radiator for communication spacecraft application. Life test report, ESA-CR(P)-1082, May 1978, Dornier-System GmbH., Friedrichshafen, W. Germany.
112. Dornier-system. Design and development of a gas controlled heat pipe radiator for communication spacecraft application. Life test report, ESA-CR(P)-1082, May 1978, European Space Agency, Paris, France.
113. Dreher, E. Theory and practice of rotary heat regenerators. (In German.) *Klima Kalte Ing.*, 1978, **6**(2), 63–69.
114. Dreitser, G. A., Markovskii, P. M. and Evdokimov, V. D. Effect of hydrodynamic instability on heat transfer during the flow of gas and liquid in tubes. (In Russian.) *Vestsi Akad. Navuk BSSR, Ser. Fiz.-Energ. Navuk*, 1978, (3), 111–19.

115. DUGAN, C. F., VAN NOSTRAND, W. L. and HALUSKA, J. L. How antifoulants reduce the energy consumption of refineries, *Chem. Eng. Prog.*, 1978, **74**(5), 53–57.
116. DUNN, A. and JAMES, R. W. An experimental rig to study the condensation of pure and mixed refrigerants, *Refrig. Air Cond.*, Dec. 1977, **80**(957), 39, 41, 44 and 46–47.
117. DUSCHA, W. Refrigeration systems in processing technology. (In German.) *Chem.-Tech.* (*Heidelberg*), 1978, **7**(8), 323–28.
118. DYER, J. R. Natural–convective flow through a vertical duct with a restricted entry, *Int. J. Heat Mass Trans.*, 1978, **21**(10), 1341–54.

119. ECKERT, E. R. G., SPARROW, E. M., GOLDSTEIN, R. J., *et al.* Heat transfer—a review of 1977 literature, *Int. J. Heat Mass Trans.*, 1978, **21**(10), 1269–98.
120. EDWARDS, T. C. A new air conditioning, refrigeration and heat pump cycle, *ASHRAE Trans.*, 1978, Part 2, **84**(2), 150–72.
121. EICKE, K. Industrial heat pumps. (In German.) *Klima Kalte Ing.*, 1978, **6**(7/8), 275–78.
122. ELECTRICAL REVIEW. Air conditioning with heat reclaim, *Elect. Rev.*, Jan. 1978, **202**(2), 46.
123. ELECTRICAL REVIEW INTERNATIONAL. Heat pipes go to great lengths, *Elect. Rev. Int.*, Feb. 1979, **204**(6), 8.
124. EMERSON, W. H. Plate heat exchangers. In: *An Introduction to Industrial Heat Exchangers*, Birniehill Institute, 18–19 Sept., 1973. Course Paper No. 5. East Kilbride, Glasgow: National Engineering Laboratory, 1973.
125. ENGELHORN, H. R. and WICKENHAEUSER, G. Heat transfer during bubble and film boiling of halogen coolants in a large pressure and temperature range. (In German.) *Goldschmidt Informiert*, 1977, **39,** 27–32.
126. ENGINEERING. Fluidised bed combustion and heat treatment, *Engineering*, 1978, **218**(12), Tech. File No. 60.
127. ENGINEERING SCIENCES DATA UNIT. The momentum change component of pressure change in two-phase flow and other non-equilibrium effects, ESDU-78001, 1978, Engineering Sciences Data Unit, London, ISBN 0-85679-218-7.

128. FAN, L. T., TOJO, K. and CHANG, C. C. Modelling of shallow fluidized-bed combustion of coal particles, *IEC Process Des. Dev.*, 1979, **18**(2), 333–37.
129. FASSASSI, D. L. and DUMINIL, M. Heat transfer to halocarbons boiling on a horizontal plane plate. (In French.) *Rev. gen. Froid*, 1978, **69**(2), 103–8.
130. FAVA, J. A. and THOMAS, D. L. Use of chlorine to control OTEC biofouling, *Ocean Eng.*, 1978, **5**(4), 269–88.
131. FEARON, J. Heat from cold—energy recovery with heat pumps, *Chart. Mech. Eng.*, Sept. 1978, **25**(8), 49–53.
132. FEDKIW, P. and NEWMAN, J. Low Peclet number behavior of the transfer rate in packed beds, *Chem. Eng. Sci.*, 1978, **33**(8), 1043–48.
133. FEDKIW, P. and NEWMAN, J. Numerical calculations for the asymptotic, diffusion dominated mass-transfer coefficient in packed bed reactors, *Chem. Eng. Sci.*, 1978, **33**(11), 1563–66.
134. FEKOVICH, J. G., GRANNEMANN, G. N., MALHALINGAM, L. M., *et al.*

Degradation of heat transfer rates due to biofouling and corrosion at Keahole Point, Hawaii, COO-4041-7, 1977, Carnegie-Mellon Univ., Pittsburgh, Pa.

135. FIELD, A. A. Heat pump progress in Europe, *Heat. Vent. Eng.*, July/Aug. 1978, **52**(610), 5–8.

136. FILIPPOV, G. A., SALTANOV, G. A. and GEORGIEV, K. G. Investigation of the effect of the dispersion, concentration and disintegration of drops on the energetic and flowrate characteristics of vapor-drop flows. (In Russian.) *Inzh.-Fiz. Zh.*, Dec. 1978, **35,** 1059–65.

137. FINLAY, I. C. Air/water sprays as potential coolants for pin-finned cold plates. Final scientific technical report, May 1971, ARL-71-0087, Ohio State Univ. Research Foundation, Columbus, Ohio.

138. FINLAY, I. C. Spray flow as heat-transfer media, *Chem. Process.*, May 1971, London, 25–27 and 29.

139. FINLAY, I. C. An analysis of heat transfer during flow of an air/water mist across a heated cylinder, *Can. J. chem. Eng.*, June 1971, **49**(3), 333–39.

140. FINLAY, I. C. and SMITH, J. Response of a single/two-pass liquid–liquid heat exchanger to disturbances in flow rate, *J. mech. Eng. Sci.*, 1967, **9**(3), 211–17.

141. FLACK, R. D., JR. and WITT, C. L. Velocity measurements in two natural convection air flows using a laser velocimeter, *J. Heat Trans.*, May 1979, **101**(2), 256–60.

142. FLUIDS GROUP APPLIED HEAT TRANSFER DIVISION. Design of evaporators: report of a meeting at NEL, 10th January, 1967, NEL Report 329, October 1967, National Engineering Laboratory, East Kilbride, Glasgow.

143. FLUIDS GROUP APPLIED HEAT TRANSFER DIVISION. Direct-contact heat transfer: report of a meeting at NEL, 15 January, 1969, NEL Report 453, May 1970, National Engineering Laboratory, East Kilbride, Glasgow.

144. FLUIDS GROUP APPLIED HEAT TRANSFER DIVISION. Compact heat exchangers: report on a meeting at NEL, 22 October, 1969, NEL Report 482, May 1971, National Engineering Laboratory, East Kilbride, Glasgow.

145. FLUIDS GROUP APPLIED HEAT TRANSFER DIVISION. Air coolers, cooling towers and evaporative coolers: report of a meeting at NEL, 24 November, 1971, NEL Report 534, Dec. 1972, National Engineering Laboratory, East Kilbride, Glasgow.

146. FLUIDS GROUP APPLIED HEAT TRANSFER DIVISION. The heat pipe forum. Papers presented at a meeting at NEL, 18 March, 1975, NEL Report 607, January 1976, National Engineering Laboratory, East Kilbride, Glasgow.

147. FLUIDS GROUP APPLIED HEAT TRANSFER DIVISION. Steam turbine condensers: report of a meeting at NEL, 17–18 September, 1974, NEL Report 619, August 1976, National Engineering Laboratory, East Kilbride, Glasgow.

148. FOLSOM, L. R. Thermal energy storage/heat engine for highway vehicle propulsion, CONF-771037, 4–6 Oct. 1977, 13th Contractors Coord. Meeting, Dearburn, Mich.

149. FOX, U. and SCHNEIDER, W. Investigating the economy of an outside air–water heat pump with a gas boiler. (In German.) *Heiz. Luft. Haustech.*, 1978, **29**(8), 299–301.

150. FRAAS, A. P. and HOLCOMB, R. S. Atmospheric fluidized-bed combustion technology test unit for industrial cogeneration plants, *Proc. Int. Conf. Fluid. Bed Combus.*, 1977 (Pub. 1978), **5**(3), 55–71.

151. France, D. M., Carlson, R. D., Chiang, T., *et al.* Characteristics of transition boiling in sodium-heated steam generator tubes, *J. Heat Trans.*, May 1979, **101**(2), 270–75.
152. Freese, H. L. and Glover, W. B. Mechanically agitated thin-film evaporators, *Chem. Eng. Prog.*, 1979, **75**(1), 52–58.
153. Friedel, L. Pressure drop during gas/vapor–liquid mixture flow in pipes. (In German.) *VDI-Ber.*, 1977 (Pub. 1978), **315**, 104–17.
154. Fry, D. J., Adams, E. E. and Jirka, G. H. Evaluation of mixing and recirculation in generic OTEC discharge designs, COO-4683-1, 1977, Mass. Inst. of Tech., Cambridge, Mass.
155. Fujii, T., Fujii, M., Tanaka, H., *et al.* Laminar free convective heat transfer from a vertical surface with uniform heat flux to air. (In Japanese.) *Nippon Kikai Gakkai Rombunshu*, 1978, **44**(387), 3832–37.
156. Fujii, T., Nagata, T. and Shinzato, K. Condensation of water vapour and heat transfer from humid air to horizontal tubes in a bank. (In Japanese.) *Refrigeration*, Dec. 1977, **52**(602), 1059–68.
157. Fukutsuka, T., Shimogohri, K., Sato, H., *et al.* Corrosion resistance of titanium pipes in multistage flash process desalination plants. (In Japanese.) *R & D, Res. Dev.* (*Kobe Steel Ltd*), 1978, **28**(3), 76–80.
158. Furukawa, M. and Sugita, H. Condensation of steam on a horizontal single tube and on a horizontal bundle of three tubes, *Kobe Shosen Daigaku Kiyo, Dai-2-Rui*, 1978, **26**(2), 101–12.

159. Gaigalis, V. A., Asakavichyus, I. P. and Eva, V. K. A reciprocating heat pipe, *Inzh.-Fiz. Zh.*, Nov. 1978, **35**(5), 773–76. (English Trans.: *J. Eng. Phys.*, May 1979, **35**(5), 1265–68.)
160. Ganapathi, K. Estimation of heat transfer coefficients in utility steam generator components, Proc. 4th Nat. Heat Mass Transfer Conf., 1977, Sarita Prakashan, Meerut, India, 729–45.
161. Ganic, E. N. and Rohsenow, W. M. On the mechanism of liquid drop deposition in two-phase dispersed flow, *J. Heat Trans.*, May 1979, **101**(2), 288–94.
162. Gardiner, S. R. M. and Sabersky, R. H. Heat transfer in an annular gap, *Int. J. Heat Mass Trans.*, 1978, **21**(12), 1459–66.
163. Garnsey, R. Boiler corrosion and the requirement for feed- and boiler-water chemical control in nuclear steam generators, *Proc. Int. Conf. Water Chem. Nucl. React. Syst.*, 1977, BNES, 1978, London, 1–10 and 33–35.
164. Garonne, X., Guillaume, R., Papon, A., *et al.* A new method of calculating the influence of condensation in the development of thermal plumes. (In French.) *Int. J. Heat Mass Trans.*, 1979, **22**(1), 21–26.
165. Gartling, D. K. NACHOS: a finite element computer program for incompressible flow problems. Part 1. Theoretical background. SAND-77-1333, 1978, Sandia Lab., Albuquerque, N. Mex.
166. Gdalevich, L. B., Nogotov, E. F. and Fertman, V. E. Effect of wall thickness and heat conductivity on heat transfer during laminar natural convection of air in a rectangular cavity. (In Russian.) *Inzh.-Fiz. Zh.*, Dec. 1978, **35**, 1130–35.
167. Gebhart, B., Bendell, M. S. and Shaukatullah, H. Buoyancy-induced

flows adjacent to horizontal surfaces in water near its density extremum. *Int. J. Heat Mass Trans.*, 1979, **22**(1), 137–49.

168. GEL'PERIN, N. I., ZAKHARENKO, V. V. and AINSHTEIN, V. G. Factor of segregation of finely divided materials in a fluidized bed. (In Russian.) Deposited Doc. VINITI 3318-77, VINITI, 1977, Moscow, USSR.
169. GENDRIN, F. Gas-driven heat pumps. (In French.) *Chaud. Froid. Plomb.*, Oct. 1978, **32**(386), 69–79.
170. GERASIMOV, YU. F., KISEEV, V. M., MAIDANIK, YU. F., *et al.* Low temperature heat pipes with vapour injection. (In Russian.) *Inzh.-Fiz. Zh.*, 1977, **33**(4), 573–80.
171. GERICKE, B. The natural circulation in waste-heat steam generators. (In German.) *Brennst.-Warme-Kraft*, 1978, **30**(12), 459–68.
172. GILMAN, S. F. Solar energy: present and future, *ASHRAE J.*, 1978, **20**(11), 33–36.
173. GODLEWSKI, J. and LASSOTA, J. Halocarbon shell-and-tube condensers. Intensification of heat exchange. (In Polish.) *Chlodnictwo*, 1978, **13**(1), 16–19.
174. GOGONIN, I. I., DOROKHOV, A. R. and SOSUNOV, V. I. Heat transfer during film condensation of stationary vapor on a vertical surface. (In Russian.) *Inzh.-Fiz. Zh.*, 1978, **35**(6), 1050–58.
175. GOLOVNYA, V. N., ZDANOVICH, N. N. and NEMTSEV, V. A. Experimental study of the thermal and hydraulic characteristics of direct-flow regenerator–evaporator prototypes. (In Russian.) In: *Dissotsiiruyushchie Gazy Teplonositeli Rab. Tela Energ. Ustanovok*, Volume 2, 1978, A. K. Krasin (ed.), Akad. Nauk Belorusskoi SSR, Inst. Teplo Massoobmena, Minsk, USSR.
176. GONZALEZ, J. A. Aspects of corrosion in boilers. (In Spanish.) *Montagjes e Instalaciones*, Dec. 1978, **8**(94), 43–49.
177. GORDIAN ASSOCIATES, INC. Heat pump technology: a survey of technical developments, market prospects and research needs, HCP/M2121-01, June 1978, Gordian Associates, Inc., New York, N.Y.
178. GRAEFEN, H., HORN, E. M. and GRAMBERG, U. Corrosion. (In German.) In: *Ullmanns Encykl. Tech. Chem., 4. Aufl.*, 1978, Volume 15, E. Bartholome, E. Biekert and H. Hellmann (eds.), Verlag Chem., Weinheim, Germany, 1–59.
179. GRAKOVICH, L. P. Two-component heat pipes. (In Russian.) In: *Nizkotemp. Tepl. Truby Poristye Teploobmenniki*, 1977, L. L. Vasil'ev (ed.), Akad. Nauk BSSR, Inst. Teplo- Massoobmena im. A. V. Lykova, Minsk, USSR, 82–91.
180. GRANT, I. D. R. Condenser performance—the effect of different arrangements for venting non-condensing gases, *Brit. Chem. Eng.*, 1969, **14**(12), 1709–11.
181. GRANT, I. D. R. and CHISHOLM, D. Two-phase flow on the shell-side of a segmentally baffled shell-and-tube heat exchanger, *J. Heat Trans.*, 1979, **101**(1), 38–42.
182. GRAY, D. E. *Proceedings of Heat Pump Workshop, Cosener's House, Abingdon, 26–27 Sept. 1977, Organised by the SRC, Rutherford Laboratory*, TRC T78 3749 (Science Research Council RL-77-145/C), Dec. 1977, Dept. of Industry, Techlink Unit, Technology Reports Centre, Orpington, Kent.
183. GREEN, H. E. Finned aluminium tubes keep the heat down, *Elec. Rev.*, 11 Aug. 1978, **203**(6), 22–23.

184. GREK, F. Z. Physical nature of the Prandtl number during heat transfer in a fluidized bed. (In Russian.) *Zh. Prikl. Khim.*, 1978, **51**(8), 1807–9.
185. GREWAL, N. S. and SAXENA, S. C. Investigations of heat transfer from immersed tubes in a fluidized bed. Proc. 4th Nat. Heat Mass Trans. Conf., 1977, Sarita Prakashan, Meerut, India, 53–58.
186. GRIFFITH, P., AVEDISIAN, C. T. and WALKUSH, J. P. Counterflow critical heat flux, *AIChE Symp. Ser.*, 1978, **74**(174), 149–55.
187. GRIGOR'EV, V. A., KLIMENKO, A. V. and PAVLOV, YU. M. Determination of the boundary of the self-modelling zone of bubble boiling in relation to the thickness of the heating wall. (In Russian.) *Tr. Mosk. Energ. Inst.*, 1977, **347,** 42–53.
188. GROCHAL, B. and TARASEWICZ, W. Study of heat penetration during the condensation of nitrogen tetroxide on a horizontal smooth or finned tube. (In Polish.) *Pr. Inst. Masz. Przeplyw., Pol. Akad. Nauk*, 1978, **76,** 69–75.
189. GRONSKII, R. K. and MAKLAKOVA, V. P. Corrosion protection of copper alloys during chemical cleaning of heat exchangers. (In Russian.) *Vodopodgotovka, Vod. Rezhim i Khimkontrol na Parosil Ustanovkakh*, 1978, (6), 69–73.
190. GRUNDSELL, B. Hard deposits in boilers reduce energy efficiency. (In Swedish.) *VVS(Sweden)*, 1979, **50**(1), 81 and 86.
191. GRUSH, W. H. and WHITE, J. R. Prediction of LOFT core fluid conditions during blowdown and refill, *Trans. Am. Nucl. Soc.*, 1978, **30,** 395–96.
192. GUNN, G. Designing efficient shell boiler plant, *Heat. Air Condit. J.*, Jan. 1979, **49**(564), 12–18.
193. GUPTA, J. P. *Heat Exchanger Design—a Practical Look*, 1979, C. S. Enterprises, New Delhi, India.

194. HAGNER, B. Comparison of different types of ventilation heat recovery equipment. (In Finnish.) *LVI.*, 1978, **30**(5), 41–44.
195. HAIGH, W. S., MARGOLIS, S. G. and RICE, R. E. Evaluation of the RELAP 4/MOD 6 thermal-hydraulic code, CONF-781022-22, 16 Oct. 1978, Meeting on Nuclear Power Reactor Safety, Brussels.
196. HALLGREEN, K. Control and regulation of solar heating plant. (In Danish.) *VVS (Denmark)*, 1978, **14**(9), 21–30.
197. HALOZAN, H. and GRAZ, E. Design and layout of residential solar heating installations with integral heat pumps. (In German.) *Technik. am Bau.*, 1978, (12), 999–1002.
198. HALS, F. A., GANNON, R. E., BECKER, F. E., *et al.* High temperature air preheaters for open cycle MHD energy conversion systems, *AIChE Symp. Ser.*, 1978, **74**(174), 302–19.
199. HANDROCK, W. Application of gas-fired heat pump technology. (In German.) *Kalte*, 1978, **31**(9), 382–97.
200. HANEL, B. Approximate calculation of free convection flow with mass transfer at a vertical wall. (In German.) *Luft- Kaeltetech.*, 1978, **14**(4), 198–202.
201. HARDY, C. W. E. Efficient use of energy—combustion control, *Heat. Vent. Eng.*, Feb. 1979, **53**(615), 16–21.
202. HARRIS, S. D. Spray cooling of heated cylinders, *AIChE Symp. Ser.*, 1978, **74**(174), 67–72.

203. HEATING AND AIR CONDITIONING JOURNAL. Package boiler survey 1978. New companies, new models. *Heat. Air Condit. J.*, Aug. 1978, **48**(559), 28–82.
204. HEIBURG, O. Gas engine-driven heat pump installation—design and operating experience. (In German.) *Klima Kalte Ing.*, 1978, **6**(7/8), 261–64.
205. HEMMINGS, J. W. and KERN, J. The non-adiabatic calorimeter problem and its application to transfer processes in suspensions of solids, *Int. J. Heat Mass Trans.*, 1979, **22**(1), 99–109.
206. HENDRICKS, R. C. and PAPELL, S. S. Estimating surface temperature in forced convection nucleate boiling—a simplified method, *Adv. Cryog. Eng.*, 1977 (Pub. 1978), **23,** 301–4.
207. HENRY, J. Auxiliary heat exchangers in power plant. In: *An Introduction to Industrial Heat Exchangers.* Birniehill Institute, 18–19 Sept., 1973. Course Paper No. 8. East Kilbride, Glasgow: National Engineering Laboratory, Birniehill Institute, 1973.
208. HENRY, R. E. and FAUSKE, H. K. Nucleation processes in large scale vapor explosions, *J. Heat Trans.*, May 1979, **101**(2), 280–87.
209. HIRAMA, T., YUMIYAMA, M., TOMITA, M., *et al.* Lateral thermal diffusivities in packed fluidized beds of horizontal flow type. (In Japanese.) *Kagaku Kogaku Rombunshu*, 1977, **3**(4), 344–48.
210. HIRAOKA, S., YAMADA, I., MIZOGUCHI, K., *et al.* Heat transfer of non-Newtonian fluid in anchor agitated vessel, *Nagoya Kogyo Daigaku Gakuho*, 1977 (Pub. 1978), **29,** 549–52.
211. HIWADA, M., TAGUCHI, T., MABUCHI, I., *et al.* Fluid flow and heat transfer around two different circular cylinders in crossflow. (In Japanese.) *Nippon Kikai Gakkai Rombunshu*, 1978, **44**(385), 3134–44.
212. HOFMANN, A. and SCHNAPPER, C. Heat transport with helium in rotating thermosiphon loops, Proc. 7th Int. Cryog. Eng. Conf., 1978, 378–85.
213. HOLMBERG, R. B. Combined heat and mass transfer in regenerators with hygroscopic materials, *J. Heat Trans.*, May 1979, **101**(2), 205–10.
214. HONEYWELL, INC. Solar heating and cooling systems design and development, NASA-CR-150786, July 1978, Honeywell, Inc., Energy Resource Center, Minneapolis, Minnesota.
215. HOTANI, S. Heat transferring characteristics of refrigerants. (In Japanese.) *Reito*, 1977, **52**(600), 963–72.
216. HUBER, D. A. Evaluation of a pressurized–fluidized bed combustion (PFBC) combined cycle power plant conceptual design. Task 1 report. Commercial plant requirements definition, FE-2371-25, Feb. 1978, Burns and Roe Industrial Services Corp., Paramus, N. J.

217. IDAHO NATIONAL ENGINEERING LABORATORY. Code development and analysis program. RELAP 4/MOD 7 (Version 2): user's manual, CDAP-TR-78-036, Aug. 1978, Idaho National Engineering Laboratory, Idaho Falls, Idaho.
218. IL'ICHEVA, S. I. Quick-response electric laboratory furnace with programmed temperature control. (In Russian.) *Sb. Tr.-Gos. Vses. Nauchno-Issled. Inst. Stroit. Mater. Konstr.*, 1977, **37,** 184–87.
219. INOUE, A. and BANKOFF, S. G. Destabilization of film boiling due to arrival of a pressure shock. 1 Experimental. In: *Top. Two-Phase Heat Transfer Flow,*

ASME Winter Annual Meeting, S. G. Bankoff (ed.), 1978, ASME, New York, N.Y., 77–88.

220. INSTRUMENT SOCIETY OF AMERICA. American National Standard for Temperature Measurement Thermocouples, ANSI-MC96.1, 1975, Instrument Society of America, Pittsburgh, Pa.
221. INTERNATIONAL INSTITUTE OF REFRIGERATION. Heat exchangers. Air conditioning. Heat pumps. Meeting of Commissions B1 (thermodynamics and transport processes), B2 (refrigerating machinery) and E1 (air conditioning) of the I.I.R., Belgrade, Yugoslavia, 16–18 Nov. 1977. 1977-4, International Institute of Refrigeration, Paris, France.
222. INTERTECHNOLOGY CORPORATION. Economic and technical feasibility analysis of Ocean Thermal Energy Conversion, ITC-0058A, NSF/PRA-77-SP-0335, 17 Feb. 1977, Intertechnology Corp., Warrenton, Pa.
223. IOUP, G. E. *Proceedings of the 4th annual conference on Ocean Thermal Energy Conversion, New Orleans, LA, 22 March 1977*, CONF-770331, July 1977, Energy Res. and Dev. Admin., Div. of Solar Energy, Washington, D.C.
224. IRIE, F., MATSHUSHITA, T., TAKEO, M., *et al.* Heat transfer to helium in the near-critical region, *Adv. Cryog. Eng.*, 1977 (Pub. 1978), **23,** 326–32.
225. ISACHENKO, V. P. Heat transfer in condensation. (In Russian.) In: *Energiya*, 1977, Moscow, USSR.
226. ISHIGURO, R., KUMADA, T. and SUGIYAMA, K. Heat transfer around a circular cylinder in liquid sodium crossflow. (In Japanese.) *Hokkaido Daigaku Kogakubu Kenkyu Hokoku*, 1978, **86,** 133–47.
227. ISTVANFY, G., GESZTI, P. and KOSARAS, F. New equipment for measuring the surface heat transfer coefficient. (In Hungarian.) *Elektrotechnika*, 1978, **71**(1), 9–16.
228. IVANOVA, V. S. Heat transfer from finned air conditioners during frost formation. (In Russian.) *Kholod. Tekh.*, 1978, (11), 57–61.
229. IZUMI, R., *et al.* Condensing heat transfer of a refrigerant in the presence of air and oil. (In Japanese.) *Refrigeration*, Jan. 1978, **53**(603), 11–17.

230. JACKSON, W. D. MHD electrical power generation: program status report, *DOE Symp. Ser.*, 1978, **46** (Sci. Probl. Coal Util., 1977, CONF-770509), 113–33.
231. JALURIA, Y. A study of the vapor region in a condensation thermal processing system, Proc. 4th Nat. Heat Mass Transfer Conf., 1977, Sarita Prakashan, Meerut, India, 461–68.
232. JASICKI, K. A. Heat transfer and flow resistance in adjustable arrangements of finned tubes. (In Polish.) *Ciep. Ogrz. Went.*, 1978, **10**(12), 322–26.
233. JENG, D. R., DEWITT, K. J. and LEE, M. H. Forced convection over rotating bodies with non-uniform surface temperature, *Int. J. Heat Mass Trans.*, 1979, **22**(1), 89–98.
234. JIRKA, G. H., JOHNSON, R. P., FRY, D. J., *et al.* Ocean Thermal Energy Conversion Plants: experimental and analytical study of mixing and recirculation, OSP-83746. R77-38. COO-2909-3, Sept. 1977, Massachusetts Inst. of Tech., Ralph M. Parsons Lab. for Water Resources and Hydrodynamics, Cambridge, Mass.

235. JOHNSON, M. W. Secondary flow in rotating bends, *J. Eng. Power*, Oct. 1978, **100,** 553–60.
236. JOHNSON, W. W. and JONES, M. C. Measurements of axial heat transport in helium II with forced convection, *Adv. Cryog. Eng.*, 1977 (Pub. 1978), **23,** 363–70.
237. JUSTIN, B. Performance testing of solar collectors, *Sun World*, Aug. 1978, **2**(3), 66–71.

238. KALININ, E. K., DREITSER, G. A. and KHASAEV, N. O. Experimental study of improvement of heat transfer during film boiling of cryogenic liquids in tubes. (In Russian.) *Izv. Akad. Nauk SSSR, Energ. Transp.*, 1978, (6), 106–15.
239. KALISCHER, P. Bivalent heat pumps—heating method, economy, practice. (In German.) *Technik am Bau.*, 1978, (8), 641–45.
240. KAMOTANI, Y., OSTRACH, S. and MIAO, H. Convective heat transfer augmentation in thermal entrance regions by means of thermal instability, *J. Heat Trans.*, May 1979, **101**(2), 222–26.
241. KANNAPPAN, D. Combustion aspects of MHD power generation, *Proc. 4th Nat. Conf. I.C. Engines Combust., 1977, Volume 2, C3*, 1978, Coll. Eng., Madras, India, 93–97.
242. KAST, W. and KLAN, H. Heat transfer with free convection of air in heated, vertical channels and room heating bodies. (In German.) *Gesundh-ing.*, 1978, **99**(7), 185–91.
243. KATO, K., OSAWA, Y., SAWADA, A., *et al.* Fluidized conditions for two-stage packed fluidized bed connected with a downcomer. (In Japanese.) *Kagaku Kogaku Rombunshu*, 1977, **3**(1), 83–87.
244. KATO, K., MAENO, M. and HASUKO, S. The lateral effective thermal conductivity in a packed fluidized bed, *J. Chem. Eng. Jpn.*, 1979, **12**(1), 38–42.
245. KATTO, Y. A generalized correlation of critical heat flux for the forced convection boiling in vertical uniformly heated round tubes, *Int. J. Heat Mass Trans.*, 1978, **21**(12), 1527–42.
246. KATTO, Y. and SHIMIZU, M. Upper limit of CHF in the saturated forced convection boiling on a heated disk with a small impinging jet, *J. Heat Trans.*, May 1979, **101**(2), 265–69.
247. KATZOFF, S. A siphon method of determining resistivities of thin heat-pipe wicks, NASA-TP-1304. L-12266, Oct. 1978, NASA, Langley Research Center, Hampton, Va.
248. KENNEY, C. D. D. and FELDMAN, K. T., JR. Heat pipe life tests at temperatures up to 400 °C, In: *Intersociety 13th Energy Conversion Eng. Conf., San Diego, Calif. 20–25 Aug. 1978. Proceedings, Volume 2*, 1978, Society of Automotive Engineers, Inc., Warrendale, Pa., 1056–59.
249. KHALIL, K. H. (ed.) *Flow, Mixing and Heat Transfer in Furnaces*, 1978, Pergamon Press, Oxford, ISBN 0-08-022695-7.
250. KHAN, A. R., RICHARDSON, J. F. and SHAKIRI, K. J. Heat transfer between a fluidized bed and a small immersed surface, In: *Fluid., Proc. 2nd Eng. Found. Conf., 1978*, J. F. Davidson and D. L. Keairns (eds.), 1978, Cambridge Univ. Press, London.
251. KIELC, E., SZEIBEL, S. and STRZESZEWSKI, R. Effective solution of the heat

conductivity equation of low Fourier and Biot numbers. (In Polish.) *Pr. Inst. Met. Niezelaz.*, 1978, **7**(1), 12–16.

252. KIM, H. D. A study on the natural convection flow between vertical plates with constant heat flux, *Yongnam Taehakkyo Kongop Kisul Yonguso Yongu Pogo*, 1978, **6**(1), 33–39.

253. KIRILLOV, V. A. and OGARKOV, B. L. Study of heat and mass transfer in a three-phase, fixed catalyst bed. (In Russian.) *Hung. J. Ind. Chem.*, 1978, **6**(2), 175–86.

254. KIYOTA, M. and URAKAWA, K. Numerical calculation methods for heat conduction with phase change. (In Japanese.) *Tokushima Daigaku Kogakubu Kenkyu Hokoku*, 1978, **23,** 107–14.

255. KOBIYAMA, M., TANIGUCHI, H. and SAITO, T. Numerical analysis of combined radiative and convective heat transfer. Part 1. Effect of two-dimensional radiative transfer between isothermal parallel plates. (In Japanese.) *Nippon Kikai Gukkai Rombunshu*, 1978, **44**(385), 3124–33.

256. KOELTECHN. KLIMAATR. Energy economiser in heating, cooling and ventilation systems. (In Dutch.) *Koeltechn. Klimaatr.*, 1978, **71**(3), 34–36.

257. KOKOTT, D. Thermodynamic studies on a wet/dry cell cooling tower. (In German.) *Chem.-Ing.-Tech.*, 1978, **50**(7), 546–47.

258. KOLYKHAN, L. I. and BATISHCHEVA, T. M. Heat exchange and hydraulic resistance during laminar condensation of nitrogen tetroxide in a vertical pipe, (English Trans.: *J. Eng. Phys.*, Nov. 1978, **34**(5), 544–49.) *Int. Aerospace Abstr.*, 1979, **19**(9–34), Abstract A79-25217.

259. KORMANYOS, K. Improved heat exchanger materials, Highway Vehicle Systems Summary Report 13, March 1978, Dept. of Energy, Transportation Energy Conservation Div., Washington, D.C., 225–38.

260. KOZIOL, K. and KONIECZNY, J. Heat transfer in scraped-surface heat exchangers. (In Polish.) *Inz. Apar. Chem.*, 1978, **17**(1), 1–5.

261. KRAUSE, W. B. An investigation of the heat transfer mechanisms around horizontal bare and finned heat exchange tubes submerged in an air fluidized bed of uniformly sized particles, Thesis, Univ. Nebraska, Lincoln, Nebraska, 1978, Order No. 7901936, 1979, Univ. Microfilms Int., Ann Arbor, Mich.

262. KRAUSE, H. H., VAUGHAN, D. A. and SEXTON, R. W., *et al.* Erosion–corrosion effects on boiler tube metals in a multisolids fluidized-bed coal combustor, *J. Eng. Power*, 1979, **101**(1), 1–8.

263. KRAVCHUN, S. N. and FILIPPOV, L. P. Radiative–conductive heat transfer in the temperature-wave regime. (In Russian.) *Inzh.-Fiz. Zh.*, Dec. 1978, **35,** 1027–33.

264. KRAVTSOV, G. M., PROTOD'YAKONOV, I. O. and TSIBAROV, V. A. Kinetic theory of a fluidized bed. (In Russian.) *Teor. Osn. Khim. Tekhnol.*, 1978, **12**(5), 716–21.

265. KRIVOSHEEV, V. F. and MIROSHNICHENKO, L. E. Some characteristics of hydrogen sulfide corrosion of carbon steel in methanol aqueous solutions. (In Russian.) *Korroz. Zashch. Truboprovodov, Skvazhin, Gazopromysl. Gazopererab. Oborudovaniya*, 1978, (6), 8–12.

266. KRUPICZKA, R., BYLICA, I., WALCZYK, H., *et al.* Study of the non-uniformity of heat transfer in shell-and-tube exchangers with various designs of the intertubular space. Part 1. Smooth tubes. (In Polish.) *Inz. Apar. Chem.*, 1978, **17**(5), 1–6.

267. Kumada, T., Ishiguro, R., Sato, T., *et al.* Natural evaporation of sodium with mist formation. Part 1. Measurement of evaporation rates and comparison of values against theoretical predictions, *J. Heat Trans.*, May 1979, **101**(2), 306–12.
268. Kuplenov, N. I. Calculation of spray-type heat exchangers. (In Russian.) *Kholod. Tekhn.*, 1978, **55**(12), 17–20.
269. Kuppers, P. State of the art and developments in heat pump research. (In German.) *Klima Kalte Ing.*, 1978, **6**(7/8), 265–69.
270. Kutateladze, S. S., Gogonin, I. I., Dorokhov, A. P., *et al.* Heat transfer during condensation on a horizontal bundle of tubes. (In Russian.) *Teploperedacha pri Kipenii i Kondensatsii*, 1978, 39–59.
271. Kydd, P. H. Composite tube heat exchangers, *AIChE Symp. Ser.*, 1978, **74**(174), 320–26.

272. Langeheinecke, K. Terminology in heat pump engineering. (In German.) *Klima Kalte Ing.*, 1978, **6**(1), 23–28.
273. Lavochnik, A. I. and Shvartsman, E. I. Boiling of refrigerating agents R-11 and R-142 and their binary mixtures in a large volume. (In Russian.) *Kholod. Tekh.*, 1978, (11), 20–22.
274. Lawaetz, H. Solar heating systems with heat pumps. (In Danish.) *VVS (Denmark)*, 1978, **14**(10), 25–28.
275. Ledinegg, M. Static pressure with two-phase flow. (In German.) *Brennst. Warme Kraft*, 1978, **30**(10), 400–6.
276. Leidenfrost, W. and Korenic, B. Analysis of evaporative cooling and enhancement of condenser efficiency and of coefficient of performance, *Warme- und Stoffubertragung*, 1979, **12**(1), 5–23.
277. Leishman, J. M. The thermal design and performance of feedheaters using superheated steam, NEL Report 467, September 1970, National Engineering Laboratory, East Kilbride, Glasgow.
278. Leishman, J. M., Vaughan, V. E., Cotchin, C. D. and Woodhead, E. The experimental performance of a desuperheating type feedheater, NEL Report 329, July 1969, National Engineering Laboratory, East Kilbride, Glasgow.
279. Lemmon, E. C. and Mackay, D. B. LOFCON-LOFT condenser program, LTR-115-28, 26 May 1978, Nat. Eng. Lab., Idaho Falls, Idaho.
280. Leon, D. M., White, M. D. and Hedrick, R. A. PWR blowdown heat transfer separate-effects program: thermal-hydraulic test facility experimental data report for test 151, ORNL/NUREG/TM-188, 1978, Oak Ridge Nat. Lab., Oak Ridge, Tenn.
281. Leon, I. W., Claypoole, G. and Reed, R. R. Multicell fluidized-bed boiler design, construction, and test program, FE-1237/M-78/64, 1978, Pope Evans and Robbins, Inc., New York, N.Y.
282. Leong, S. S. and De Vahl Davis, G. Natural convection in a horizontal cylinder, Report 1979/FMT/5, Feb. 1979, Univ. of New South Wales, Kensington, Australia.
283. Lewis, L. G. Ocean Thermal Energy Conversion (OTEC) power plant instrumentation and measurement, CONF-780550-5, Solar seminar on testing solar energy materials and systems, 22–24 May 1978, Gaithersburg, Md.

284. LIBURDY, J. A., GROFF, E. G. and FAETH, G. M. Structure of a turbulent thermal plume rising along an isothermal wall, *J. Heat Trans.*, May 1979 **101**(2), 249–55.
285. LIENHARD, J. H. and HASAN, M. M. On predicting boiling burnout with the mechanical energy stability criterion, *J. Heat Trans.*, May 1979, **101**(2), 276–79.
286. LINDSTROM, Y. and RYTKONEN, V. Condenser heat recovery in refrigeration equipment. (In Finnish.) *LVI.*, 1978, **30**(4), 30–33.
287. LIPMAN, M. Save: condense ammonia with R-12, *Hydrocarb. Process.*, 1979, **58**(1), 153–56.
288. LISIN, V. V., CHEPURNENKO, V. P. and GOGOL, N. I. Comparison of ammonia condensers with air cooling. (In Russian.) *Kholod. Tekh. Tekhnol.*, 1978, **26,** 86–89.
289. LOCKHEED MISSILES AND SPACE COMPANY, INC. Ocean Thermal Energy Conversion (OTEC) development of an apparatus to measure biofouling effects on heat exchangers in ocean environments. Design report., LMSC-D-562046, Oct. 1977, Lockheed Missiles and Space Co., Inc., Ocean Systems Div., Sunnyvale, Calif.
290. LORENZI, A. and MUZZIO, A. Two-phase flow measurement with sharp-edged orifices. Experimental results. (In Italian.) *Termotecnica*, 1978, **32**(10), 527–34.
291. LORENZINI, E. and ORLANDELLI, C. M. Experimental study on forced convection in annular ducts. Part 1. (In Italian.) *Tec. Ital.*, 1977, **42**(3), 127–35.
292. LORENZINI, E. and SPIGA, M. Temperature distribution in a hot spot of a cylindrical fuel rod. (In Italian.) *Termotecnica*, 1978, **32**(10), 544–51.
293. LUBISCHIK, U. Solar-supplemented heat pump installation. (In German.) *Clima. Commer. Internat.*, 1978, **12**(10), 38–39.
294. LUCCI, A. D. Advanced coal-fuelled combustor/heat exchanger technology study. Quarterly Technical Progress Report No. 3, 1 Sept.–30 Nov. 1977, FE-2612-10, 15 Dec. 1977, Rockwell Internat., Rocketdyne Div., Canoga Park, Calif.
295. LYCZKOWSKI, R. W., GIDASPOW, D. and GALLOWAY, T. R. Hydrodynamic modelling of fluidized-bed gasifiers and combustors, UCID-17759, 19 April 1978, Calif. Univ., Lawrence Livermore Lab., Livermore, Calif.

296. MAA, J. R. Drop size distribution and heat flux of dropwise condensation, *Chem. Eng. J. (Lausanne)*, 1978, **16**(3), 171–76.
297. MCCREIGHT, L. R. Heat exchanger materials based on lithium and magnesium aluminum silicates, Highway Vehicles Systems Summary Report 13, March 1978, Dept. of Energy, Transportation Energy Conservation Div., 224–25.
298. MCENANEY, B. and SMITH, D. C. Corrosion of a cast iron boiler in a model central heating system, *Corros. Sci.*, 1978, **18**(7), 591–603.
299. MACHBITZ, M., BUDIMAN, B., ROBERTS, Y. Y., *et al.* Radiation heat transfer in a 19-pin sodium-voided bundle, *Trans. Am. Nucl. Soc.*, 1978, **30,** 416–17.
300. MCKENZIE, R. T. B. Refrigeration theory: the vapour-compression cycle, *Austral. Refrig. Air Cond. Heat.*, 1977, **31**(11), 29–30, 35–38; (12), 27–30, 35–38; 1978, **32**(1), 29–30, 35–38.

301. MADER, H. An experimental heat pump installation with latent heat storage for residential heating. (In German.) *Elektrowarme. Int. A.*, Nov. 1978, **36**(6), 319–24.

302. MAGRINI, U. and NANNEI, E. Effect of thickness and thermal characteristics of walls on the coefficient limits in nucleate boiling. (In Italian.) *Termotecnica*, 1978, **32**(9, Supplement), 22–29.

303. MAKHORIN, K. E. and YANATA, I. I. Temperature distribution in the fluidized bed of an inert heat transfer agent. (In Russian.) *Khim. Tekhnol.* (*Kiev*), 1978, **101**, 39–41.

304. MAKHORIN, K. E., PIKASHOV, V. S. and KUCHIN, G. P. Measuring particle temperature and emissivity in a high temperature fluidized bed, In: *Fluid., Proc. 2nd Eng. Found. Conf.. 1978*, J. F. Davidson and D. L. Keairns (eds.), 1978, Cambridge Univ. Press, London, 93–97.

305. MANN, D. L. and STYHR, K. H. Preliminary evaluation of tubular ceramic heat exchanger materials, TID-28279, 8 Dec. 1977, Ai Research Mfq. Co., Torrance, Calif.

306. MANVI, R. Development of a direct-contact heat exchanger. Phase 1 study report, NASA-CR-157737. ME-78-NSG-7229, 30 Sept. 1978, Calif. State Univ., Dept. of Mech. Eng., Los Angeles, Calif.

307. MARSHALL, R. H. Natural convection effects in rectangular enclosures containing a phase-change material, In: *Therm. Storage Heat Transfer Sol. Energy Syst., Winter Annual Meeting ASME, 1978*, F. Kreith, R. Boehm and J. Mitchell (eds.), ASME, 1978, New York, N.Y., 61–69.

308. MARTIN, H. Low Peclet number particle-to-fluid heat and mass transfer in packed beds, *Chem. Eng. Sci.*, 1978, **33**(7), 913–19.

309. MARTIN, H. Velocity profiles of steady laminar two-phase flow in horizontal tubes. Comments, *Chem. Eng. Sci.*, 1978, **33**(9), 1297–98.

310. MARTINET, J. Heat pipes and thermally anisotropic structures, DOE-tr-94, 1978, Dept. of Energy, Washington, D.C.

311. MARTO, P. J., MACKENZIE, D. K. and RIVERS, A. D. Nucleate boiling in thin liquid films, *AIChE Symp. Ser.*, 1977, **73**(164), 228–35.

312. MATSUDA, T., *et al.* A new air-source heat pump system, *ASHRAE J.*, 1978, **20**(8), 32–35.

313. MATSUURA, A., AKEHATA, T. and SHIRAI, T. Effect of latent heat on temperature profile for concurrent air–water downflow in packed beds. (In Japanese.) *Kagaku Kogaku Rombunshu*, 1977, **3**(5), 528–30.

314. MATVEEV, V. M., FILIPPOV, YU. N., SILAEV, A. F., *et al.* Investigation of heat pipes with a capillary structure made of a porous material, *High Temp.*, Sept.–Oct. 1978, **16**(5), 894–98.

315. MATVEEV, V. M., FILIPPOV, YU. N., SILAEV, A. F., *et al.* Study of heat pipe with capillary structure made from powdered material. (In Russian.) *Teplofiz. Vys. Temp.*, 1978, **16**(5), 1054–59.

316. MAZYUKEVICH, I. V. Condensation of saturated vapor from a mixture on a vertical surface. (In Russian.) *Zh. Prikl. Khim.*, 1978, **51**(6), 1304–8.

317. MEHRA, YU. R. Refrigerants. Part 2. Propylene. Refrigerant charts for propylene systems, *Chem. Eng.* (*N.Y.*), 1979, **86**(2), 131–39.

318. MEHRA, YU. R. Refrigerants. Part 3. Ethane. Charts for systems using ethane refrigerant, *Chem. Eng.* (*N.Y.*), 1979, **86**(4), 95–101.

319. MEHROTRA, A. K., ALAM, S. S. and VARSHNEY, B. S. Pool boiling of liquid mixtures, Proc. 4th Nat. Heat Mass Transfer Conf., 1977, Sarita Prakashan, Meerut, India, 397–408.
320. MERKIN, J. H. Free convection boundary layers in a saturated porous medium with lateral mass flux, *Int. J. Heat Mass Trans.*, 1978, **21**(12), 1499–1504.
321. MESSINA, A. D. Effects of precise arrays of pits on the nucleate boiling of Freon 113 from flat copper surfaces at one atmosphere pressure, Thesis, University of Missouri, Rolla, Mo. 1978, Order No. 7904328, 1979, Univ. Microfilms Int., Ann Arbor, Mich.
322. MEYER, B. A., EL-WAKIL, M. M. and MITCHELL, J. W. Natural convection heat transfer in small and moderate aspect ratio enclosures. An application to flat plate collectors, In: *Therm. Storage Heat Transfer Sol. Energy Syst., Winter Annual Meeting ASME, 1978*, F. Kreith, R. Boehm and J. Mitchell (eds.), ASME, 1978, New York, N.Y., 29–33.
323. MEYER, F. K. and MERBACH, A. E. Temperature measurement under high pressure using commercial platinum resistors, *J. Phys. E.*, 1979, **12**(3), 185–86.
324. MEYER, W. Coal fired combined power station unit with fluidised-bed combustion. (In German.) *Brennst.-Warme-Kraft*, 1978, **30**(11), 430–32.
325. MICHELS, H. T., KIRK, W. W. and TUTHILL, A. H. The role of corrosion and fouling in steam condenser performance, *Nucl. Energy* (*Br. Nucl. Energy Soc.*), 1978, **17**(4), 335–42.
326. MINCHEV, P. M., KALNINS, T., KHRISTOV, KH. D., *et al.* Optimization of active elements of induction liquid metal MHD-pumps with a transverse closed up magnetic flow. (In Russian.) *Magn. Gidrodin.*, 1978, (3), 71–74.
327. MISHINA, L. V. and SEREBRYANYI, G. Z. Analytical solutions of temperature distribution in laminar flow of non-equilibrium mixtures in a flat channel with constant wall temperature. (In Russian.) In: *Dissotsiiruyushchie Gazy Teplonositeli Rab. Tela Energ. Ustanovok*, Volume 2, 1976, A. K. Krasin (ed.), Akad. Nauk BSSR, Inst. Teplo-Massoobmena, Minsk, USSR, 72–80.
328. MIYAMOTO, M. and SUMIKAWA, J. Effect of wall heat conduction on laminar free convection from a vertical flat plate. (In Japanese.) *Yamaguchi Daigaku Kogakubu Kenkyu Hokoku*, 1978, **29**(1), 161–70.
329. MIZUSHINA, T., ITO, R., YAMASHITA, S., *et al.* Film condensation of superheated pure vapor in a vertical tube. (In Japanese.) *Kagaku Kogaku Rombunshu*, 1977, **3**(3), 296–302.
330. MOALEM-MARON, D. and ZIJL, W. Growth, condensation, and departure of small and large vapor bubbles in pure and binary systems, *Chem. Eng. Sci.*, 1978, **33**(10), 1339–46.
331. MOCHIZUKI, S. and SHIRATORI, T. Condensation heat transfer within a circular tube under centrifugal acceleration. (In Japanese.) *Nippon Kikai Gakkai Rombunshu*, 1978, **44**(383), 2420–28.
332. MOKHTARZADEH, M. R. and EL-SHIRBINA, A. A. A theoretical analysis of evaporating droplets in an immiscible liquid, *Int. J. Heat Mass Trans.*, 1979, **22**(1), 27–38.
333. MONTEIRO, J. L. F. and SADDY, M. A study on segregation in gas fluidized beds. (In Portuguese.) *Cientifica*, 1977, (Special issue), 103–9.
334. MORI, Y. Heat transfer bibliography—Japanese works, *Int. J. Heat Mass Trans.*, 1979, **22**(1), 151–56.

335. MOSHFEGHIAN, M. Fluid flow and heat transfer in U-bends, Thesis, Oklahoma State University, Stillwater, Okla., 1978, Order No. 7903711, 1979, Univ. Microfilms, Ann Arbor, Mich.
336. MOTWANI, D. G. and STHAPAK, B. K. Correlation of local heat transfer and pressure drop in a natural circulation horizontal tube evaporator, Proc. 4th Nat. Heat Mass Transfer Conf., 1977, Sarita Prakashan, Meerut, India, 345–52.
337. MUHLBAIER, D. R. and HARRIS, S. D. Calculation of two-phase transient flow in low pressure systems, DP-MS-77-26 CONF-780620-1, 1978, National Technical Information Service, Springfield, Va.
338. MULLER, H. The heat pump, a means of saving energy. (In German.) *Elektrowarmetechn. Ausbau*, 1977, **35**(4), A222–29.
339. MUROYAMA, K., HASHIMOTO, K. and TOMITA, T. Heat transfer from wall in gas–liquid concurrent packed beds, *Heat Transfer-Jap. Res.*, 1978, **7**(1), 87–93.
340. MURRAY, I. Direct-contact heat exchangers. In: *An Introduction to Industrial Heat Exchangers*, Birniehill Institute, 18–19 Sept., 1973. Course Paper No. 6. East Kilbride, Glasgow: National Engineering Laboratory, 1973.
341. MURRAY, I. Speed up exchanger design with STEM, *Heat. Vent. Eng.*, Feb. 1977, **51**(595), 5–8.
342. MYASNIKOV, V. P. Non-steady-state motions in a fluidized bed. (In Russian.) *Dokl. Akad. Nauk SSSR*, 1978, **243**(1), 163–66. (Chem. Tech.)

343. NA, T. Y. and CHIOU, J. P. Turbulent heat transfer for pipe flow with uniform heat generation, *Warme- und Stoffubertrangung*, 1979, **12**(1), 55–58.
344. NAG, P. K. and RAO, D. N. Heat transfer from gas–solid suspensions, Proc. 4th Nat. Heat Mass Transfer Conf., 1977, Sarita Prakashan, Meerut, India, 235–46.
345. NAKAMURA, M., HIOKI, K., TAKAHASHI, A., *et al.* Fluid flow characteristics of the three-phase fluidized bed with perforated plates. (In Japanese.) *Kagaku Kogaku Rombunshu*, 1978, **4**(5), 473–77.
346. NAKAMURA, H., MATSUURA, A., KIWAKI, J., *et al.* The effect of variable viscosity on laminar flow and heat transfer in rectangular ducts, *J. Chem. Eng. Jpn.*, Feb. 1979, **12**(1), 14–18.
347. NAKAZATOMI, M., TAHARA, K. and HAMASHIN, M. Forced convective heat transfer to air–water bubble flow. Part 1. Experimental results on the relationship between radial void fraction distribution and heat transfer coefficient. (In Japanese.) *Ube Kogyo Koto Semmon Gakko Kenkyu Hokoku*, 1978, **24,** 21–35.
348. NARAYANAN, R. Free surface convection in cylindrical geometries, Thesis, Illinois Inst. Technol., Chicago, Ill., 1978. Order No. 7902991, 1979, Univ. Microfilms Int., Ann Arbor, Mich.
349. NATIONAL ENGINEERING LABORATORY. Design of shell-and-tube heat exchangers, NEL, 24–26 April 1978, Paper No. 3, East Kilbride, Glasgow: National Engineering Laboratory, 1978.
350. NATIONAL RESEARCH DEVELOPMENT CORPORATION. Pressurized fluidized-bed combustion. Annual report, July 1976–June 1977. Draft, HCP/T1511-43/D, Sept. 1977, Nat. Res. Dev. Corp., London.

351. NECHAEV, YU. G., MALASHIKHIN, K. V., MIKHAL'CHUK, E. M., *et al.* Method of calculating mass transfer coefficients for rotary-film contact devices. (In Russian.) *Izv. Vyssh. Uchebn. Zaved., Khim. Khim. Tekhnol.*, 1978, **21**(7), 1059–64.
352. NEIZVESTNYI, A. I. Experimental determination of the coefficient of water condensation from the evaporation and growth rates of micron-size drops. (In Russian.) *Dokl. Akad. Nauk SSSR*, 1978, **243**(3), 626–29. (Phys.)
353. NELSON, M. E. and BOCK, A. E. Dimensional analysis of Ocean Thermal Energy Conversion heat exchangers. Final report, 1 July 1976–30 June 1977, USNA-EPRD-33. AD A054 985, 30 June 1977, Naval Acad., Energy-Environmental Study Group, Annapolis, Md.
354. NESTERENKO, V. B., TVERKOVKIN, B. E. and TUSHIN, N. N. Heat and mass transfer during the turbulent flow of nitrogen oxide (N_2O_4) in axisymmetric channels. (In Russian.) *Vestsi Akad. Navuk BSSR, Ser. Fiz.-Energy. Navuk*, 1978, (3), 76–82.
355. NEW MEXICO UNIVERSITY. Heat pipe technology: a bibliography with abstracts, 1978. Quarterly Report, NTISUB/D/022, 1978, New Mexico Univ., Technol. Application Center, Albuquerque, N. Mex.
356. NEWBY, R. A. and KEAIRNS, D. L. Fluidized-bed heat transport between parallel, horizontal tube-bundles, In: *Fluid., Proc. 2nd Eng. Found. Conf., 1978*, J. F. Davidson and D. L. Keairns (eds.), 1978, Cambridge Univ. Press, London, 320–26.
357. NEWTON, D. S. Corrosion protection and materials, *Anti-Corrosion*, 1978, **25**(8), 10–12.
358. NGUYEN, H. V., WHITEHEAD, A. B. and POTTER, O. E. Gas back mixing in large fluidized beds, In: *Fluid., Proc. 2nd Eng. Found. Conf., 1978*, J. F. Davidson and D. L. Keairns (eds.), 1978, Cambridge Univ. Press, London, 140–45.
359. NGUYEN, T. V., MACLAINE-CROSS, I. L. and DE VAHL DAVIS, G. Combined forced and free convection between parallel plates, Report 1979/FMT/4, Feb. 1979, Univ. of New South Wales, Kensington, Australia.
360. NICHOLS, M. C. and GREEN, R. M. Direct-contact heat exchange for latent heat-of-fusion energy storage systems. In: *Alternative Energy Sources*, (*Proc. Miami Int. Conf.*), 1977, Volume 9, T. N. Veziroglu (ed.), 1978, Hemisphere, Washington, D.C.
361. NIINO, M., KASHINSKII, O. N. and ODNORAL, V. P. Study of the nucleate regime of the flow of a gas–liquid mixture in a vertical pipe. (In Russian.) *Inzh.-Fiz. Zh.*, Dec. 1978, **35,** 1044–49.
362. NIJAGUNA, B. T. and KRISHNAMURTHY, M. T. Heat transfer in concentric rotating cylinders, Proc. 4th Nat. Heat Mass Transfer Conf., 1977, Sarita Prakashan, Meerut, India, 119–26.
363. NISHIKAWA, K. and FUJITA, Y. Heat transfer in nucleate boiling of Freons. (In Japanese.) *Reito*, 1978, **53**(607), 389–401.
364. NISHIMATSU, A., TAJIMA, O., TSUJI, H., *et al.* Pool boiling heat transfer to aqueous lithium bromide solutions under vacuum state. 2. Effects of heating surface roughness. (In Japanese.) *Reito*, 1978, **53**(607), 381–88.
365. NISHIMURA, M., NAKAO, K., HIRABAYASHI, Y., *et al.* Influence of the radiative property of a boundary wall on the effect of infrared internal heating. (In Japanese.) *Kagaku Kogaku Rombunshu*, 1978, **4**(5), 547–49.

366. NOVOZHILOV, V. N. Sulfuric acid drying of air in apparatus with rising flow. (In Russian.) *Khim. Prom-st. (Moscow)*, 1978, (8), 627–28.

367. OCEAN DATA SYSTEMS, INC. OTEC thermal resource report for Florida east coast, TID-27948, Oct. 1977, Ocean Data Systems, Inc., Monterey, Calif.
368. OETIKER, N. K., RAO, R. B. S. and REID, M. B. Corrosion problems in cooling towers—their prevention in practice, Prepr. Pap., Annu. Conf.—Australasian Corrosion Association, 1976, Volume 16 (Corros.-Probl., Prev., Pract.), Paper No. 23.
369. OGATA, H. and NAKAYAMA, W. Boiling and free convection heat transfer to helium in centrifugal acceleration fields. (In Japanese.) *Reito*, 1978, **53**(610), 733–41.
370. OHNO, M. A simple designing method of a condenser in cooler. (In Japanese.) *Kagaku Sochi*, 1978, **20**(3), 26–36.
371. OKADA, K. and OHNO, M. Practical designing method of the heat exchangers of plate type. (In Japanese.) *Kagaku Sochi*, 1978, **20**(6), 11–19.
372. ORNATSKII, A. P., SEMENA, M. G. and TIMOFEEV, V. I. Maximum heat flux in planar metal-fiber wicks under conditions typical of heat pipes, *Inzh.-Fiz. Zh.*, Nov. 1978, **35**(5), 782–88.
373. OSINSKII, V. P. and SEMENOV, M. M. Method for design calculation of a heat exchanger with a directionally moving fluidized bed of a finely divided material and heat-transfer surfaces submerged in the bed. (In Russian.) *Tr. Vses. Nauchno-Issled. Konstr. Inst. Khim. Mashinostr.*, 1976, **75,** 138–47.
374. OSTRACH, S. and RAGHAVAN, C. Effect of stabilizing thermal gradients on natural convection in rectangular enclosures, *J. Heat Trans.*, May 1979, **101**(2), 238–43.
375. OZISIK, M. N. and UZZELL, J. C., JR. Exact solution for freezing in cylindrical symmetry with extended freezing temperature range, *J. Heat Trans.*, May 1979, **101**(2), 331–34.

376. PERELETOV, I. I., SHOPSHIN, M. F., NOVOSEL'TSEV, V. N., *et al.* Optimal design of a heat-exchanger reactor in a system for regenerative heat utilization. (In Russian.) *Tr. Mosk. Energ. Inst.*, 1977, **332,** 98–104.
377. PERRY, E. J. Heat pumps—the future, *Heat. Air Condit. J.*, Sept. 1978, **48**(559), 16–18.
378. PETERSON, A. C. and WINTERS, L. Rod thermal response during isothermal loss-of-coolant experiments, *Trans. Am. Nucl. Soc.*, 1978, **30,** 409.
379. PINKHASIK, M. S., MIRONOV, V. D., ZAKHARKO, YU. A., *et al.* Control of the fuel combustion process in the combustion chamber of a magnetohydrodynamic generator. (In Russian.) *Teploenergetika*, 1978, (5), 41–44.
380. POHLMANN, H. J., MAIER, H. R. and KRAUTH, A. Recuperative ceramic heat exchanger gas turbine application, *Proc. Br. Ceram. Soc.*, 1978, **26** (*Mech. Eng. Prop. Appl. Ceram.*), 31–40.
381. POPE, W. L., PINES, H. S., SILVESTER, L. F., *et al.* Optimizations of geothermal cycle shell-and-tube exchangers of various configurations with variable fluid properties and site-specific fouling, LBL-7039, March 1978, University of California, Berkeley, Calif.
382. POSTLETHWAITE, A. W. and SLUYTER, M. M. MHD heat transfer problems—an overview, *Mech. Eng.*, 1978, **100**(3), 32–39.

383. POTTER, D. S. Use of embossed and dimpled plate heat exchange jacketing on process vessels, *Process Eng.*, April 1968.
384. POZ. M. YA., SENATOVA, V. L. and GRANOVSKII, V. L. Regenerative heat exchangers and heat exchangers with intermediate heat-transfer agents. (In Russian.) *Vodosnabzh. Sanit. Tekh.*, 1978, (12), 14–17.
385. PRASAD, R., ALAM, S. S. and VARSHNEY, B. S. Pool boiling of aqueous salt solution, Proc. 4th Nat. Heat Mass Transfer Conf., 1977, Sarita Prakashan, Meerut, India, 383–96.
386. PRITCHARD, A. B. Fluid bed combustion systems, *Energy World*, Aug./Sept. 1978, (51), 16–18.
387. PRITCHARD, A. M., PEAKALL, K. A. and SMART, E. A miniature water loop for boiler corrosion and deposition studies, In: *Water Chem. Nucl. React. Syst., Proc. Int. Conf.. 1977*, BNES, 1978, London, 139–49.
388. PRON'KO, V. G. and BULANOVA, L. B. Experimental investigation of the thermodynamic crisis of film boiling, *J. Eng. Phys.*, Nov. 1978, **34**(5), 534–39.
389. PYLILO, L. E. Study of parameters of a low-temperature heat pipe using ammonia in relation to the amount of the working liquid. (In Russian.) In: *Nizkotemp. Tepl. Truby Poristye Teploobmenniki*, 1977, L. L. Vasil'ev (ed.), Akad. Nauk BSSR, Inst. Teplo- Massoobmena im. A. V. Lykova, 1977, Minsk, USSR, 92–96.

390. QUON, C. A study of penetrative convection in rotating fluid, *J. Heat Trans.*, May 1979, **101**(2), 261–64.

391. RADKE, M. Energy-saving evaporation-heat pumps for distillation processes. (In German.) *Chem.-Anlagen Verfahren*, 1978, (12), 55–56.
392. RAJAKUMAR, A. and KRISHNASWAMY, P. R. Experimental frequency domain dynamics of heat pipes, *Int. J. Heat Mass Trans.*, 1978, **21**(10), 1333–40.
393. RANKEN, W. A. Heat pipe applications workshop report, LA-7229-C. CONF-7710144-1, 20 Oct. 1977, Heat Pipe Applications Workshop, Los Alamos, New Mexico.
394. RAO, G. V. and NARAYANASWAMI, R. Optimization of a conducting cooling fin with a heat sink using optimality criterion, *Int. J. Solids Structures*, 1978, **14**(10), 787–93.
395. RASHID, K. and KUWABARA, S. Natural convection in a confined region between two concentric square ducts, *J. Phys. Soc. Jpn.*, 1978, **45**(6), 2014–20.
396. RASHID, K. and KUWABARA, S. Natural convection in a confined region between two horizontal ducts for variable fluid properties, *J. Phys. Soc. Jpn.*, 1978, **45**(6), 2021–29.
397. RAVINDRAN, P. V. and RAO, C. D. Heat exchanger dynamics and parameters dependence on flow rates, *Indian J. Technol.*, 1978, **16**(1), 1–4.
398. RAZELOS, P. An analytic solution to the electric analog simulation of the regenerative heat exchanger with time-varying fluid inlet temperatures, *Warme- und Stoffubertrangung*, 1979, **12**(1), 59–71.
399. REAY, D. A. Heat pipe heat exchangers—current status and development, *Heat Vent. Eng.*, Jan. 1979, **53**(615), 8–14.
400. REAY, D. A. *Heat pumps. Design, construction and applications*, June 1979, Pergamon Press, Oxford, ISBN 0-08-022716-3.

401. Regef, A. Operating costs of dual flow systems. Comparative study of several alternatives using heat exchangers and heat pumps. (In French.) *Froid Clim.*, Jan.–Feb. 1978, (293), 72–77.
402. Rennebeck, K. Anti-freezing protection for air conditioning and ventilation installations by regenerative recovery of latent and sensible heat energy. (In German.) *Klima Kalte Ing.*, 1978, **6**(11), 417–26.
403. Rigot, G. Continuous and cyclic operation of compression refrigerating plant. (In French.) *Rev. prat. Froid. Condit. Air*, Jan. 1978, **31**(436), 51–60; Feb. 1978, (438), 39–50.
404. Rios, G. and Gibert, H. Mechanical stirring of fluidized beds: potential improvements in the control of heat transfer, In: *Fluid., Proc. 2nd Eng. Found. Conf., 1978*, J. F. Davidson and D. L. Keairns (eds.), 1978, Cambridge Univ. Press, London. 357–61.
405. Roehm, H-J. Calculation of condensation of multicomponent vapor mixtures. (In German.) *Chem.-Ing.-Tech.*, 1978, **50**(12), 963.
406. Roetzel, W. Calculation of conduction and radiation with film boiling from a flat plate. (In German.) *Warme- und Stoffubertragung*, 1979, **12**(1), 1–4.
407. Rosard, D. D. Working fluids and turbines for OTEC power systems, In: *Fluids Eng. Adv. Energy Syst., Winter Annual Meet. ASME, 1978*, C. H. Marston (ed.), 1978, ASME, New York, N.Y., 229–46.

408. Saluja, S. N. and James, R. W. Heat transfer and pressure drop characteristics of binary refrigerant mixtures, *Refrig. Air Cond.*, Feb. 1978, **81**(959), 32–35.
409. Saluja, S. N. and James, R. W. Operating characteristics of a mixed refrigerant vapour compression system, *Refrig. Air Cond.*, March 1978, **81**(960), 80, 84, 86, 113.
410. Sanders, R. C. and Mueller, G. E. Loss-of-coolant accidents in small compact nuclear reactors, *Nucl. Technol.*, 1979, **42**(3), 289–96.
411. Sawyer, R. F., Schefer, R. W., Ganji, A. R., *et al.* Laboratory studies of lean combustion, NASA-CP-2021, Aircr. Engine Emiss., 1977, N78-11063, National Aeronautics and Space Admin., 1977, Washington, D.C. 417–36.
412. Saxena, P. K. and Shah, K. S. Thermal insulation of buried pipes with polyurethane foam, Proc. 4th Nat. Heat Mass Transfer Conf., 1977, Sarita Prakashan, Meerut, India, 31–38.
413. Schley, U. Advances in the area of temperature measurement. (In German.) *PTB-Mitt.*, 1979, **89**(1), 13–21.
414. Schmitt-Thomas, K. G., Meisel, H. and Birkmayer, E. Protection layer formation and corrosion in crude gasoline condensers. (In German.) *Erdoel Kohle, Erdgas, Petrochem.*, 1978, **31**(9), 412–16.
415. Schmitz, U. Air type evaporator for heat pump systems. (In German.) *Klima Kalte Ing.*, 1978, **6**(7/8), 253–56.
416. Scholz, W. H. Coiled tubular heat exchangers, Conference on Heat Transfer and the Design and Operation of Heat Exchangers, University of the Witwatersrand, 17–19 April 1974.
417. Sedler, B. and Mikielewicz, J. Simplified analytical model of a flow boiling crisis. (In Polish.) *Pr. Inst. Masz. Przeplyw., Pol. Akad. Nauk*, 1978, **76,** 3–10.

418. Seetharam, T. R. and Sharma, G. K. Numerical predictions of free convective heat transfer from vertical surfaces to fluids in the near-critical region, Proc. 4th Nat. Heat Mass Transfer Conf., 1977, Sarita Prakashan, Meerut, India, 273–79.
419. Seetharam, T. R. and Sharma, G. K. Free convective heat transfer to fluids in the near-critical region from vertical surfaces with uniform heat flux, *Int. J. Heat Mass Trans.*, 1979, **22**(1), 13–20.
420. Seki, K. and Ichinoseki, K. Study on pitting corrosion of copper tubes for heat exchangers. (In Japanese.) *Shindo Gijutsu Kenkyu Kaishi*, 1977, **16**(1), 33–40.
421. Seki, M., Kawamura, H. and Sanokawa, K. Natural convection of mercury in a magnetic field parallel to the gravity, *J. Heat Trans.*, May 1979, **101**(2), 227–32.
422. Semena, M. G. Maximum heat-transfer capacity of vertical two-phase thermosiphons. (In Russian.) *Inzh.-Fiz. Zh.*, 1978, **35**(3), 397–403.
423. Semena, M. G. and Kiselev, Yu. F. Study of heat transfer in the heat supply zone of two-phase thermosiphons at low degrees of filling. (In Russian.) *Inzh.-Fiz. Zh.*, 1978, **35**(4), 600–5.
424. Semena, M. G. and Nishchik, A. P. Structure parameters of metal-fiber heat pipe wicks, *Inzh.-Fiz. Zh.*, Nov. 1978, **35**(5), 777–81. (English Trans.: *J. Eng. Phys.*, May 1979, **35**(5), 1268–72.)
425. Seppanen, O. and Heino, R. Calculating heat recovery in air conditioning installations. (In Finnish.) *L.V.I., Fl.*, 1978, **30**(1), 28–31.
426. Shamsundar, N. and Srinivasan, R. Analysis of energy storage by phase change with an array of cylindrical tubes, In: *Therm., Storage Heat Transfer Sol. Energy Syst., Winter Annual Meeting ASME, 1978*, F. Kreith, R. Boehm and J. Mitchell (eds.), 1978, ASME, New York, N.Y., 35–40.
427. Sharma, B. I. Analytical study of laminar-free convection in oils, Proc. 4th Nat. Heat Mass Transfer Conf., 1977, Sarita Prakashan, Meerut, India, 267–71.
428. Sharma, C. P., Agrawal, K. N. and Singh, G. Boiling heat transfer of subcooled water in forced convection, Proc. 4th Nat. Heat Mass Transfer Conf., 1977, Sarita Prakashan, Meerut, India, 363–81.
429. Sharma, P. R., Kumar, P., Gupta, S. C., *et al.* Nucleate pool boiling of liquids on horizontal cylinder at subatmospheric pressure, Proc. 4th Nat. Heat Mass Transfer Conf., 1977, Sarita Prakashan, Meerut, India, 423–33.
430. Sharma, P. R., Kumar, P., Gupta, S. C., *et al.* Application of typical existing correlations to the boiling heat transfer data at sub-atmospheric pressure, Proc. 4th Nat. Heat Mass Transfer Conf., 1977, Sarita Prakashan, Meerut, India, 435–50.
431. Shcherbakov, V. K. and Kovalenko, L. V. Study of temperature distribution in the walls of interfin channels during boiling of a moving subcooled liquid in them. (In Russian.) *Inzh.-Fiz. Zh.*, 1978, **35**(3), 404–9.
432. Sheth, V. M., Shah, B. M. and Gomkale, S. D. Pneumatic drying: heat and mass transfer studies, Proc. 4th Nat. Heat Mass Transfer Conf., 1977, Sarita Prakashan, Meerut, India, 567–85.
433. Shibayama, S., Katsuta, M., Suzuki, K., *et al.* Study on boiling heat transfer in thin liquid film. Part 1. In the case of pure water and aqueous solutions of

surface active agents as working liquids. (In Japanese.) *Nippon Kikai Gakkai Rombunshu*, 1978, **44**(383), 2429–38.

434. SHVETSOV, N. A. and MIKHAILOV, V. I. Evaluation of the effect of the incompleteness of solution evaporation on the efficiency of a lithium bromide absorption refrigerating machine. (In Russian.) *Issled. Kholodil'n. Mashin.*, 1978, 152–57.
435. SILVER, R. S. *Steam Plant Aspects of Sea Water Distillation*, 1978, Mechanical Engineering Publications Ltd, London, ISBN 0-85298-385-9.
436. SINGH, M. and RAJVANSHI, S. C. Heat transfer between two parallel porous disks, *Proc. Nat. Acad. Sci., India, Sect. A*, 1978, **48**(1), 13–19.
437. SINHA, M. P., PUROHIT, N. K. and MITRA, A. K. Heat transfer in single-phase tube flow with jet mixing: an analytical approach, *Indian Chem. Eng.*, 1978, **20**(2), 38–43.
438. SIT, S. P. and GRACE, J. R. Interphase mass transfer in an aggregative fluidized bed, *Chem. Eng. Sci.*, 1978, **33**(8), 1115–22.
439. SLOBODYANIK, I. P. Improvement in mass transfer in a regularly rotating two-phase flow. (In Russian.) *Teor. Osn. Khim. Tekhnol.*, 1978, **12**(5), 643–48.
440. SMITH, E. M. Effectiveness—NTU relationships for tubular exchangers, *Int. J. Heat and Fluid Flow*, March 1979, **1**(1), 43–46.
441. SMITH, M. F. Boiler corrosion (citations from the NTIS Data Base), NTIS/PS-78/0993, 1978, National Technical Information Service, Springfield, Va.
442. SMITH, M. F. Boiler corrosion. Volume 2, 1975–1976 (citations from the Engineering Index Data Base), NTIS/PS-78/0994, 1978, National Technical Information Service, Springfield, Va.
443. SMITH, M. F. Boiler corrosion. Volume 3, 1977–Aug. 1978 (citations from the Engineering Index Data Base), NTIS/PS/78/0995, 1978, National Technical Information Service, Springfield, Va.
444. SMITH, R. V. and AZZOPARDI, B. J. Summary of reported droplet size distribution data in dispersed two-phase flow, NUREG-CR-0476, Oct. 1978, Wichita State Univ., Kan.
445. SNYDER, T., BENTLEY, J., GIEBLER, M., *et al.* Advanced wet–dry cooling tower concept performance prediction, MIT-EL-77-002 (V.1). COO-2500-2 (V.1), Jan. 1977, Massachusetts Inst. of Tech., Energy Lab., Cambridge, Mass.
446. SNYDER, T., BENTLEY, J., GIEBLER, M., *et al.* Advanced wet–dry cooling tower concept cross-flow tests, MIT-EL-77-002 (V.2). COO-2500-2 (V.2), Jan. 1977, Massachusetts Inst. of Tech., Energy Lab., Cambridge, Mass.
447. SOLAND, J. G., *et al.* Performance ranking of plate-fin heat exchanger surfaces, *Trans. ASME*, Aug. 1978, **100**(3), 514–19.
448. SOLOPENKOV, K. N. and LYAPIN, V. V. Ratio between vapor and liquid phases in uniformly heated pipes of vertical thermosiphon evaporators. (In Russian.) In: *Massoobmennye Teploobmennye Protsessy Khim. Tekhnol.*, 1976, B. N. Basargin (ed.), 1976, Yaroslavskii Politekh. Inst., Yaroslavl, USSR, 152–64.
449. SOLOPENKOV, K. N., LYAPIN, V. V., KLYKOVA, L. I., *et al.* Coefficients of heat transfer to binary azeotropic mixtures boiling in vertical pipes. (In Russian.) In: *Massoobmennye Teploobmennye Protsessy Khim. Tekhnol.*, 1976, B. N. Basargin (ed.), 1976, Yaroslavskii Politekh. Inst., Yaroslavl, USSR, 146–51.
450. SOLOUKHIN, R. L. and MARTYNENKO, O. G. Heat and mass transfer bibliography—Soviet works, *Int. J. Heat Mass Trans.*, 1979, **22**(1), 157–66.

451. SONNICHSEN, J. C., JR. Calculation of evaporative loss coefficients for thermal power plants, HEDL-TME-78-33, June 1978, Hanford Eng. Dev. Lab., Richland, Wash.
452. SOTNIKOV, A. G., *et al.* Heat exchange with non-steady thermal processes in an air conditioned room. (In Russian.) *Kholod. Tekhn.*, 1978, **55**(7), 30–32.
453. SOUNDALGEKAR, V. M., POHANERKAR, S. G. and WAVRE, P. D. On unsteady free convective flow of water at 4 °C past an infinite vertical plate with variable suction, Proc. 4th Nat. Heat Mass Transfer Conf., 1977, Sarita Prakashan, Meerut, India, 301–6.
454. SPARKS, D. T., STANLEY, C. J. and MCCARDELL, R. K. Film boiling behaviour in a nine-rod cluster, *Trans. Am. Nucl. Soc.*, 1978, **30**, 404–5.
455. SPARROW, E. M. and CESS, R. D. *McGraw-Hill Series in Thermal and Fluids Engineering. Radiation Heat Transfer*, 1978, McGraw-Hill, London.
456. SPEDDING, P. L. and NGUYEN, V. T. Bubble rise and liquid content in horizontal and inclined tubes, *Chem. Eng. Sci.*, 1978, **33**(8), 987–94.
457. SPYRIDONOS, A. V. Experimental study of a heat pipe with an active porous medium. In: *Alternative Energy Sources*, (*Proc. Miami Int.* (*Conf.*), 1977, Volume 2, T. N. Veziroglu (ed.) Hemisphere, Washington, D.C. 619–27.
458. STENZEL, A. Design considerations for heat pump installations, *Refrig. Air Condit.*, Dec. 1978, **81**(969), 32–38.
459. STEPANOV, V. S., PETROV, V. A. and BITYUKOV, V. K. Radiative–conductive heat transfer in a plane layer of selective medium with semi-transparent boundaries. (In Russian.) *Teplofiz. Vys. Temp.*, 1978, **16**(6), 1277–84.
460. STEPHAN, K. Heat transfer with change of phase in engineering processes. (In German.) *VDI-Ber.*, 1977 (Pub. 1978), **315**, 72–79.
461. STEPHAN, K. and PREUSSER, P. Heat transfer and maximum heat flux density in the vessel boiling of binary and ternary liquid mixtures. (In German.) *Chem.-Ing.-Tech.*, 1979, **51**(1), 37.
462. STONE AND WEBSTER ENGINEERING CORPORATION. Technical notes for the conceptual design for an atmospheric fluidized-bed direct combustion power generating plant, HCP/T2583-01/1, 1978, Stone and Webster Eng. Corp., New York, N.Y.
463. STRAZZA, N., BRENNAN, P. J. and NGUYEN, N. H. Copper/water axially grooved heat pipes for RTG applications, *Intersociety 13th Energy Conversion Eng. Conf., San Diego, Calif.. 20–25 Aug. 1978. Proceedings, Volume 2*, Society of Automotive Engineers, Inc., 1978, Warrendale, Pa. 1707–11.
464. STREL'TSOV, YU. A. Calculation of a reactor with a fluidized bed and a fixed packing under external diffusion resistance. (In Russian.) Deposited Doc. VINITI 1453-77, 1977, VINITI, Moscow, USSR.
465. STRONG, D. T. G. The fossil-fuel fired heat pump, *Build. Serv. Environ. Eng.*, Oct. 1978, **1**(2), 18–19.
466. STRUCK, W., *et al.* Heat generation with diesel or gas-engine heat pumps. (In German.) *Klima Kalte Ing.*, 1978, **6**(7/8), 279–84.
467. STUKALENKO, A. K. and BARANENKO, A. V. Thermodynamic analysis of cycles of absorption refrigerating machines using lithium bromide-water and choline chloride-water solutions. (In Russian.) *Issled. Kholodil'n. Mashin.*, 1978, 34–40.

468. STYRIKOVICH, M. A., POLONSKII, V. S. and ZUIKOV, A. S. Effect of iron oxide deposits on the temperature of steam-generating channels in the supercritical region. (In Russian.) *Teplofiz. Vys. Temp.*, 1978, **16**(6), 1314–16.
469. SUGIYAMA, S. and HASATANI, M. Unsteady heat conduction accompanied by an endothermic solid reaction, *Mem. Fac. Eng., Nagoya Univ.*, 1978, **30**(1), 110–51.
470. SUN, T. H. and PRAGER, R. C. Analysis and application of the heat pipe exchanger, *Intersociety 13th Energy Conversion Eng. Conf., San Diego, Calif.. 20–25 Aug. 1978. Proceedings, Volume 2*, 1978, Society of Automotive Engineers, Inc., Warrendale, Pa.
471. SURINOV, YU. A. Iterative-zonal method for studying and calculating radiative heat exchange in an absorbing and scattering medium. (In Russian.) *Izv. Sib. Otd. Akad. Nauk SSSR, Ser. Tekh. Nauk*, 1978, (2), 106–25.
472. SYSKA, A. J. New concepts in burner design, *Adv. Instrum.*, 1978, **33**(2), 121–30.
473. SZARGUT, J. and WANDRASZ, J. Calculating the radiant exchange in a closed chamber by the method of radiation bands as compared with the method of the average radiation parameter. (In German.) *Brennst. Warme Kraft.*, 1978, **30**(8), 324–29.
474. TAGHAVI-TAFRESHI, K., DHIR, V. K. and CATTON, I. Thermal and hydrodynamic phenomena associated with melting of a horizontal substrate placed beneath a heavier immiscible liquid, *J. Heat Trans.*, May 1979, **101**(2), 318–25.
475. TAIROV, I. P. Effect of cooling conditions on heat transfer during vapor condensation. (In Russian.) *Nekotorye Probl. Teplo- i Massoobmena*, 1978, 120–22.
476. TAIROV, I. P. Study of heat transfer during vapor condensation inside vertical channels. (In Russian.) *Nekotorye Probl. Teplo- i Massoobmena*, 1978, 123–26.
477. TEICHEL, H. The calculation of stationary two-phase flows in rod bundles. (In German.) *Brennst.-Warme-Kraft*, 1978, **30**(11), 435–40.
478. THIELBAHR, W. H. Heat exchanger technology needs for conservation research and technology, ERDA, NWC-TM-2930, Dec. 1976, Naval Weapons Center, China Lake, Calif.
479. THOME, J. R. and BALD, W. B. Nucleate pool boiling in cryogenic binary mixtures, *Proc. 7th Int. Cryog. Eng. Conf., 1978*, 4–7 July 1978, ICEC, London, 523–30.
480. THURLOW, R. M. Protection of boiler internals from corrosion—future trends, Prepr. Pap., Annu. Conf.—Australasian Corrosion Association, 1976, Volume 16, (Corros.-Probl., Prev., Pract.), Paper No. 33.
481. TOLUBINSKII, V. I., ANTONENKO, V. A., OSTROVSKII, YU. N., *et al.* Drop carry-over phenomenon in liquid evaporation from capillary structures, *Lett. Heat Mass Trans.*, 1978, **5**(6), 339–47.
482. TOMIDA, T., KUNIYOSHI, M. and OKAZAKI, T. Study on a tubular gas–liquid reactor, effects of liquid properties on the liquid-side volumetric mass transfer coefficient in upward two-phase flow, *Bull. Fac. Eng., Tokushima Univ.*, 1977, 13–14.

483. TUERLINOKX, G. and GOEDSEELS, V. Energy comparison of recuperative heat exchangers with heat pumps. (In German.) *Klima Kalte Ing.*, 1978, **6**(6), 227–30.
484. TUNIK, A., BOL'SHAKOV, A. and TEHVER, J. Effect of the porous coating of the heating surface on the rate of heat transfer during the boiling of liquid dielectrics. (In Russian.) *Eesti NSV Tead. Akad. Toim., Fuus., Mat.*, 1978, **27**(3), 364–69.

485. UHLIG, H. Heat transfer and pressure loss with condensation of water vapour from humid air in cross-flow over close-finned tube coolers. (In German.) *Klima Kalte Ing.*, 1979, **7**(1), 13–19.
486. ULRICH, G. D. Investigation of the mechanism of fly-ash formation in coal-fired utility boilers. Quarterly Report, 1 Feb.–30 April 1978, FE-2205-11, 1 June 1978, New Hampshire Univ., Dept. Chem. Eng., Durham.
487. URUSOV, E. N., SAKHNENKO, V. I. and PLATONOV, V. V. Automatic control of a semicontinuous evaporation process. (In Russian.) *Khim. Prom-st. (Moscow)*, 1978, (8), 631–33.
488. USUI, M., MARTINEZ, I., QUINTANA, R., *et al.* Corrosion of reverberatory furnace tubes. (In Spanish.) *Minerales*, 1978, **33**(143), 13–17.

489. VAJRAVELU, K. and SASTRI, K. S. Laminar free convection heat transfer of a viscous incompressible heat generating fluid-flow past a vertical porous plate in the presence of free-stream oscillations. I, II. *Acta Mech.*, 1978, **31**(1–2), 71–87, 89–100.
490. VAN DER HORST, J. F. Practical investigation of an air/air heat pump. (In Dutch.) *Verwarm. Vent.*, 1978, **35**(7), 525–32.
491. VAN DER MEER, T. H. and HOOGENDOORN, C. J. Heat transfer coefficients for viscous fluids in a static mixer, *Chem. Eng. Sci.*, 1978, **33**(9), 1277–82.
492. VAN DER REE, H. Heat-driven heat pumps. (In Dutch.) *Verwarm. Vent.*, 1978, **35**(5), 379–91.
493. VAN DYKE, M., WEHAUSEN, J. V. and LUMLEY, J. L. (eds.) Annual review of fluid mechanics. Volume II, 1979 Annual Reviews. Inc., Palo Alto, Calif.
494. VAN LIER, J. J. C. The use of the term 'energy' for the evaluation of the energy transformation in electricity and/or heat generation. (In German.) *Brennst.-Warme-Kraft*, 1978, **30**(12), 475–84.
495. VAN SCIVER, S. W. and CHRISTIANSON, O. Heat transport in a long tube of helium II, *Proc. 7th Int. Cryog. Eng. Conf., 1978*, 4–7 July 1978, ICEC, London, 228–34.
496. VAROQUAUX, E. Progress in nuclear refrigeration of helium-3, *J. Phys., Colloq. (Orsay, Fr.)*, 1978, **3**(6), 1605–12.
497. VASIL'EV, L. L. Third International Conference on Heat Pipes, *Inzh.-Fiz. Zh.*, Nov. 1978, **35**(5), 935–37. (English Trans.: *J. Eng. Phys.*, May 1979, **35**(5), 1388–90.)
498. VASIL'EV, L. L. and KONEV, S. V. Study of the operation of cryogenic heat pipes. (In Russian.) In: *Probl. Teplo-Massoobmena-77, 1977*, R. I. Soloukhin (ed.), 1977, Akad. Nauk BSSR, Inst. Teplo-Massoobmena, Minsk, USSR, 32–34.

499. VASIL'EV, L. L. and KONEV, S. V. Theory of the temperature stabilization process in gas-controlled heat pipes, *J. Eng. Phys.*, Nov. 1978, **34**(5), 527–34.
500. VELICHKO, V. I. and PRONIN, V. A. Results of an experimental study of heat transfer during flow of aqueous solutions of ethyl alcohol in a pipe. (In Russian.) *Tr. Mosk. Energ. In-ta*, 1978, (364), 57–63.
501. VERKIN, B. I. and KIRICHENKO, YU. A. Soviet investigations of pool boiling of cryogenic liquids, *Proc. 7th Int. Cryog. Eng. Conf., 1978*, 4–7 July 1978, ICEC, London, 505–22.
502. VIJAYASIMHA, C. R., VASU, K. I. and KUBAIR, V. G. Experimental investigations on free convection heat transfer to liquid metals in isothermal vertical annuli, Proc. 4th Nat. Heat Mass Transfer Conf., 1977, Sarita Prakashan, Meerut, India, 291–300.
503. VILYUNOV, V. N. and DIK, I. G. Some features of heat and mass transfer in a chemically reacting turbulent flow. (In Russian.) In: *Dissotsiiruyushchie Gazy Teplonositseli Rab. Tela Energ. Ustanovok, 1976*, Volume 2, A. K. Krasin (ed.), 1976, Akad. Nauk BSSR, Inst. Teplo Massoobmena, Minsk, USSR, 46–55.
504. VLADISLAVLEV, A. P., KIRIYA, V. V. and PONOMARENKO, YU. B. Fluidized bed in a high-frequency pulsed flow. (In Russian.) *Teor. Osn. Khim. Tekhnol.*, 1978, **12**(5), 722–26.
505. VLASOV, YU. G., GONIONSKII, V. TS., LEVERASH, V. I., *et al.* Automatic control of solid-phase concentration in apparatus of a multiple-unit evaporator for sodium chloride crystallization. (In Russian.) *Vopr. Atom. Nauki i Tekhn. Ser. Opresnenie Solen. Vod.*, 1977, (10), 81–83.
506. VOLOSHKI, A. A., VURGAFT, A. V. and FROLOV, V. N. Regimes of formation of gas bubbles in a liquid layer. (In Russian.) *Inzh.-Fiz. Zh.*, Dec. 1978, **35,** 1066–71.

507. WAGNER, H. J. and TUROWSKI, R. Energy analysis of heat pumps and thermal insulation. (In German.) *Clima Commerce Internat.*, 1978, **12**(12), 35–38.
508. WAKAO, N. and FUNAZKRI, T. Effect of fluid dispersion coefficients on particle-to-fluid mass transfer coefficients in packed beds. Correlation of Sherwood numbers, *Chem. Eng. Sci.*, 1978, **33**(10), 1375–84.
509. WAKED, A. M. and MUNSON, B. R. Laminar-turbulent flow in a spherical annulus, *J. Fluids Eng.*, Sept. 1978, **100,** 281–86.
510. WEBB, R. L. A generalized procedure for the design and optimization of fluted Gregorig condensing surfaces, *J. Heat Trans.*, May 1979, **101**(2), 335–39.
511. WESLEY, D. A. Thin disk on a convectively cooled plate—application to heat flux measurement errors, *J. Heat Trans.*, May 1979, **101**(2), 346–52.
512. WILESTEN, R. The ground as a heat pump's energy source. (In Swedish.) *Varme-o-sanit-tek.*, 1978, **42**(4), 11–13.
513. WILSON, N. W. and VYAS, B. D. Velocity profiles near a vertical ice surface melting into fresh water, *J. Heat Trans.*, May 1979, **101**(2), 313–17.
514. WINNEN, D. F. Solar heat collectors and heat pumps. (In Dutch.) *Ingenieur.*, Sept. 1978, **90**(36), 669–80.
515. WOLF, S. and HOLMES, D. H. Thermal/hydraulic test results including critical heat flux conditions for a sodium-heated steam generator tube model, NEDM-14150, 1976, Gen. Electr. Co., Fast Breeder React., Sunnyvale, Calif.

516. YACONO, C., ROWE, P. N. and ANGELINO, H. An analysis of the distribution of flow between phases in a gas fluidised bed, *Chem. Eng. Sci.*, 1979, **34**(6), 789–800.
517. YADIGAROGLU, G. The reflooding phase of the LOCA in PWRs. Part 1. Core heat transfer and fluid flow, *Nuc. Saf.*, 1978, **19**(1), 20–36.
518. YAMAMOTO, J. and SHIGI, T. Low evaporation helium cryostat with a refrigerator, *Proc. 7th Int. Cryog. Eng. Conf., 1978*, 4–7 July 1978, ICEC, London, 593–98.
519. YANG, J. W. Analysis of combined convection heat transfer in rod bundles, 7 Aug. 1978, BNL-NUREG-23018. CONF-780805-1, Heat Transfer Conf., Toronto, Canada.
520. YIM, A., EPSTEIN, M., BANKOFF, S. G., *et al.* Freezing–melting heat transfer in a tube flow, *Int. J. Heat Mass Trans.*, 1978, **21**(9), 1185–96.
521. YIN, S-H., LAI, J-Y. and HUANG, J. Natural convection heat transfer in vertical cylindrical enclosures, *Proc. Nat. Sci. Counc., Repub. China*, 1978, **2**(4), 424–30.
522. YUNG, D., LORENZ, J. J. and GANIC, E. N. Vapor/liquid interaction and entrainment in shell-and-tube evaporators, ANL-OTEC-78-2, 1978, Argonne Nat. Lab., Argonne, Ill.

523. ZAKKAY, V. and MILLER, G. Heat exchanger designs for coal-fired fluidized beds, *Intersociety 13th Energy Conversion Eng. Conf., San Diego, Calif., 20–25 Aug. 1978. Proceedings, Volume 1*, 1978, Society of Automotive Engineers, Inc., Warrendale, Pa.
524. ZAV'YALOV, V. V. Characteristics of external heat transfer of a fluidized bed under self-oscillating regime. (In Russian.) *Teplo- i Massoobmena u Mnogofazn. Mnogokomponent. Sistemakh.*, 1978, 41–45.
525. ZHAVORONKOV, N. M. and MALYUSOV, V. A. Heat transfer effect on rectification kinetics, Proc. 3rd Conf. Appl. Chem., Unit Oper. Processes, 1977, Magy. Kem. Egyesulete, Budapest, Hungary, 175–79.
526. ZSCHERNIG, J. Energy saving by optimum application of heat pumps. (In German.) *St. Gebaud*, 1978, **32**(8), 228–33.

INDEX TO BIBLIOGRAPHY

INDEX